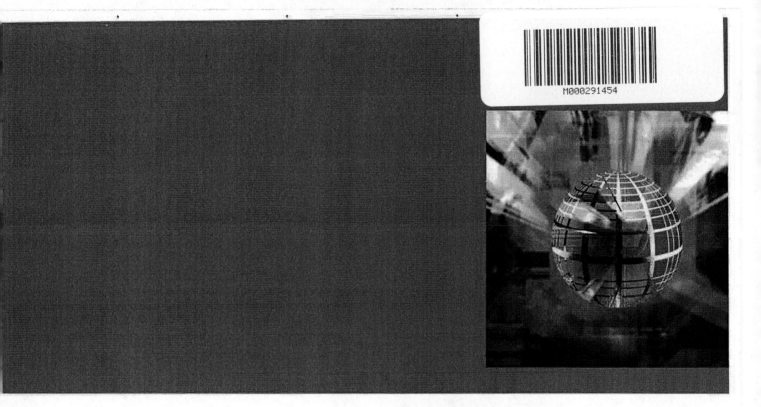

SIGNIFICANT CHANGES TO THE

A117.1 ACCESSIBILITY STANDARD

2009 EDITION

DELMAR
CENGAGE Learning

Australia • Brazil • Japan • Korea • Mexico • Singapore • Spain • United Kingdom • United States

Significant Changes to the A117.1 Accessibility Standard: 2009 Edition
International Code Council

Delmar Staff:

Vice President, Technology and Trades
  Professional Business Unit:
  Gregory L. Clayton

Product Development Manager: Ed Francis

Director of Building Trades:
  Taryn Zlatin McKenzie

Development: Dawn Jacobson

Director of Marketing: Beth A. Lutz

Marketing Manager: Marissa Maiella

Production Director: Carolyn Miller

Production Manager: Andrew Crouth

Senior Content Project Manager: Andrea Majot

Art Director: Benjamin Gleeksman

ICC Staff:

Senior Vice President, Business and Product
  Development: Mark A. Johnson

Deputy Senior Vice President, Business
  and Product Development: Hamid Naderi

Technical Director, Product Development:
  Doug Thornburg

Manager, Product and Special Sales:
  Suzane Nunes Holten

Senior Marketing Specialist: Dianna Hallmark

For product information and technology assistance, contact us at
Cengage Learning Customer & Sales Support, 1-800-354-9706

For permission to use material from this text or product, submit
all requests online at **www.cengage.com/permissions.**
Further permissions questions can be e-mailed to
permissionrequest@cengage.com

Library of Congress Control Number: 2010930947

ISBN-13: 978-1-4354-9898-3

ISBN-10: 1-4354-9898-4

ICC World Headquarters
500 New Jersey Avenue, NW
6th Floor
Washington, DC 20001-2070
Telephone: 1-888-ICC-SAFE (422-7233)
Website: **http://www.iccsafe.org**

Delmar
5 Maxwell Drive
Clifton Park, NY 12065-2919
USA

Cengage Learning is a leading provider of customized learning solutions with office locations around the globe, including Singapore, the United Kingdom, Australia, Mexico, Brazil, and Japan. Locate your local office at: **international. cengage.com/region**

Cengage Learning products are represented in Canada by Nelson Education, Ltd.

Visit us at **www.InformationDestination.com**

For more learning solutions, visit **www.cengage.com**

Notice to the Reader
Publisher does not warrant or guarantee any of the products described herein or perform any independent analysis in connection with any of the product information contained herein. Publisher does not assume, and expressly disclaims, any obligation to obtain and include information other than that provided to it by the manufacturer. The reader is expressly warned to consider and adopt all safety precautions that might be indicated by the activities described herein and to avoid all potential hazards. By following the instructions contained herein, the reader willingly assumes all risks in connection with such instructions. The publisher makes no representations or warranties of any kind, including but not limited to, the warranties of fitness for particular purpose or merchantability, nor are any such representations implied with respect to the material set forth herein, and the publisher takes no responsibility with respect to such material. The publisher shall not be liable for any special, consequential, or exemplary damages resulting, in whole or part, from the readers' use of, or reliance upon, this material.

Printed in the United States of America
1 2 3 4 5 6 7 12 11 10

Contents

This table of contents includes references from sections that are covered with a related topic elsewhere in the book. The changes and their summary could easily be overlooked since the topic is not covered in the sequential order but is instead addressed with a related change. These hidden or less apparent changes are identified by two methods in the table of contents. The "hidden" changes are indented in the table of contents and include a reference to "See page" prior to the page number. The section numbers of these hidden changes are also shown enclosed in brackets. The number following the bracket is the section numbering shown with the actual page in the book where the discussion occurs.

For example, if you look between Sections 603.2 and 603.3 you will see an entry for "(603.2, 603.2.2) 301.2 - Overlap of clearances at doors, fixtures, appliances, and elements - See page 1". This entry indicates that there was a change affecting Sections 603.2 and 603.2.2 and the "overlap of clearance" requirements that were covered in the discussion with Section 301.2 on page 1 of the book. Combining related subjects like this allowed for additional topics to be included in the book and will also help show how changes relate to similar topics in other parts of the standard.

Chapter 5.
General Site and Building Elements

No changes covered

Chapter 6.
Plumbing Elements and Facilities

Chapter 7.
Communication Elements
and Features

Chapter 11.
Recreational Facilities

Preface

The purpose of *Significant Changes to the A117.1 Standard* is to familiarize code officials, plans examiners, inspectors, design professionals, contractors, and others in the construction industry with many of the important changes in the 2009 A117.1 standard. This publication is designed to assist all users of the standard with identifying the specific code changes that have occurred and, more important, understanding the reasons behind the changes and how they will affect accessibility. It is also a valuable resource for jurisdictions in their adopting process.

Only a portion of the total number of changes to the A117.1 standard are discussed in this book. The changes selected were identified for a number of reasons, including their frequency of application, special significance, or change in application. However, the importance of those changes not included is not to be diminished. Further information on all code changes can be found on the International Code Council (ICC) website, *http://www.iccsafe.org*.

This book is organized to follow the general layout of the standard, including code sections and section number format. The table of contents, in addition to providing guidance in use of this publication, allows for quick identification of those significant changes that occur in the 2009 A117.1 standard.

Throughout the book, each change is accompanied by a photograph or an illustration to assist and enhance the reader's understanding of the specific change. A summary and a discussion of the significance of the changes are also provided. Each code change is identified by type, be it an addition, modification, clarification, or deletion.

The change to the text of the standard itself is presented in a format similar to the style utilized for submitting and reviewing proposed changes. Text deleted from the standard is shown with a strike-through, whereas new text that is added is indicated by underlining. As a result, the actual text of the 2009 standard is provided as well as a comparison with the 2003 language, so the user can easily determine changes to the specific text.

As with any code-change document, *Significant Changes to the A117.1 Standard* is best used as a study companion to the *ICC A117.1-2009*.

Because only a limited discussion of each change is provided, the standard itself should always be referenced in order to gain a more comprehensive understanding of a specific change and its application.

The commentary and opinions set forth in this text are those of the author and do not necessarily represent the official position of the ICC and are not to be considered as the opinion of the A117.1 Accredited Standards Committee. In addition, they may not represent the views of any enforcement agency, as such agencies have the sole authority to provide a review and approval process. In many cases, the explanatory material is derived from the reasoning expressed by the proponent of the change or by the A117.1 committee during its evaluation of the proposal.

Comments concerning this publication are encouraged and may be directed to the ICC at *significantchanges@iccsafe.org*.

About the A117.1 Standard

Building officials, design professionals, and others involved with building construction and accessibility recognize the need for a modern, easy-to-understand, up-to-date standard addressing the design and construction of elements that serve or are used by building occupants. The A117.1 standard, in the 2009 edition, is intended to meet these needs by providing the technical details to ensure that the buildings and facilities are accessible and usable. The A117.1 standard is kept up to date through an open development process that meets the requirements of the American National Standards Institute (ANSI) so that it can be recognized as an American National Standard (ANS). The provisions of the 2003 edition, along with those changes approved through the current development cycle, make up the 2009 edition.

The 1961 edition of the A117.1 standard presented the first criteria for accessibility to be approved as an ANS and was the result of research conducted by the University of Illinois under a grant from the Easter Seal Research Foundation. The National Easter Seal Society and the President's Committee on Employment of People with Disabilities began serving as the Secretariat, and the 1961 edition was reaffirmed in 1971.

In 1974, the U.S. Department of Housing and Urban Development joined the Secretariat and sponsored needed research, which resulted in the 1980 edition. After further revision that included a special effort to remove application criteria (scoping requirements), the 1986 edition was published. In 1987 the committee requested that the Council of American Building Officials (CABO) assume the role of the Secretariat. Central to the intent of the change in the Secretariat was the development of a standard that, when adopted as part of a building code, would be compatible with the building code and its enforcement. The 1998 edition of the A117 standard largely achieved that goal. The 2009 edition of the standard is the latest example of the A117.1 committee's effort to continue developing a standard that is compatible with the building code. In 1998 CABO became the International Code Council.

The ICC, the current Secretariat and publisher of the A117.1, was established as a nonprofit organization dedicated to developing, maintaining, and supporting a single set of comprehensive and coordinated national model building construction codes. Its mission is to provide the highest quality codes, standards, products, and services for all concerned with the safety and performance of the built environment.

The A117.1 is one of 7 standards and 14 international codes being developed and published by the ICC. This comprehensive standard establishes minimum requirements to ensure that buildings and facilities are accessible and usable. The A117.1 is available for adoption and use by jurisdictions internationally. Its use within a governmental jurisdiction is intended to be accomplished through adoption by reference, in accordance with proceedings establishing the jurisdiction's laws.

Acknowledgments

A special thank you is extended to Kim Paarlberg, Senior Staff Architect in the ICC's Codes and Standards Development department. Kim's review and input has truly helped to improve this publication. I would also like to thank A117.1 committee members Marsha Mazz, Dominic Marinelli, and Ed Roether for their assistance in securing several photos, as well as other committee members who helped explain the new requirements to me. The entire A117.1 committee should be recognized for their effort and dedication in developing the new provisions and maintaining the standard. Their efforts truly do impact and improve people's lives.

About the Author

Jay Woodward
International Code Council

Jay is a senior staff architect with the ICC's Business and Product Development department and works out of the Lenexa, Kansas, Distribution Center. His current responsibilities include serving as the Secretariat for the ICC A117.1 standard committee and assisting in the development of new ICC publications.

With more than 28 years of experience in building design, construction, code enforcement, and instruction, Jay's experience provides him with the ability to address issues of code application and design for code enforcement personnel as well as architects, designers, and contractors. Jay has previously served as the Secretariat for the ICC's *International Energy Conservation Code* and the *International Building Code's* Fire Safety Code Development committee.

A graduate of the University of Kansas and a registered architect, Jay has also worked as an architect for the Leo A. Daly Company in Omaha, Nebraska; as a building Plans Examiner for the City of Wichita, Kansas; and as a Senior Staff Architect for the International Conference of Building Officials (ICBO) prior to working for the ICC.

About the ICC

The ICC is a nonprofit membership association dedicated to protecting the health, safety, and welfare of people by creating better buildings and safer communities. The mission of the ICC is to provide the highest-quality codes, standards, products, and services for all concerned with the safety and performance of the built environment. The ICC is the publisher of the family of the International Codes® (I-Codes®), a single set of comprehensive and coordinated model codes. This unified approach to building codes enhances safety, efficiency, and affordability in the construction of buildings. The ICC is also dedicated to innovation, sustainability, and energy efficiency. In addition, the ICC Evaluation Service, an ICC subsidiary, issues Evaluation Reports for innovative products and Reports of Sustainable Attributes Verification and Evaluation (SAVE).

Headquarters: 500 New Jersey Avenue, NW, 6th Floor,
Washington, DC 20001-2070
District Offices: Birmingham, AL; Chicago, IL; Los Angeles, CA
1-888-422-7233
www.iccsafe.org

CHANGE TYPE:　Modification

CHANGE SUMMARY:　Relocates the provisions that allow clearances to overlap into the "building block" sections of Chapter 3. Allows for the elimination of redundant provisions in other locations and adds door maneuvering clearances into the permitted overlap.

301.2

Overlap of Clearances at Doors, Fixtures, Appliances, and Elements

2009 STANDARD:

301.2 Overlap. Unless otherwise specified, clear floor spaces, clearances at fixtures, maneuvering clearances at doors, and turning spaces shall be permitted to overlap.

603 Toilet and Bathing Rooms

603.2 Clearances.

~~603.2.2 Overlap.~~ ~~Clear floor spaces, clearances at fixtures, and turning spaces shall be permitted to overlap.~~

804 Kitchens and Kitchenettes

804.5 Appliances.　Where provided, kitchen appliances shall comply with Section 804.5.

804.5.1 Clear Floor Space.　A clear floor space complying with Section 305 shall be provided at each kitchen appliance. ~~Clear floor spaces are permitted to overlap.~~

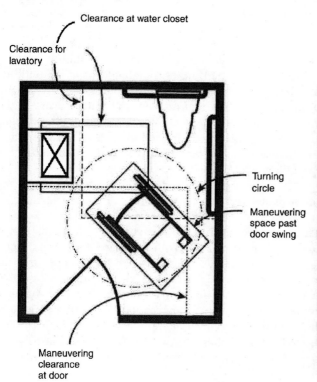

Clearance for lavatory

Clearance at water closet

Turning circle

Maneuvering space past door swing

Maneuvering clearance at door

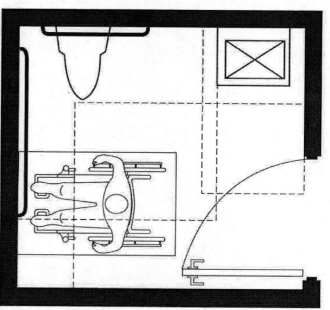

301.2 continues

301.2 continued **1003 Type A Units**

1003.11 Toilet and Bathing Facilities.

~~**1003.11.3 Overlap.** Clear floor spaces, clearances at fixtures and turning spaces are permitted to overlap.~~

1004 Type B Units

1004.11 Toilet and Bathing Facilities. Toilet and bathing fixtures shall comply with Section 1004.12.

~~**1004.11.1.3 Overlap.** Clear floor spaces shall be permitted to overlap.~~

CHANGE SIGNIFICANCE: Relocating the provisions to the building block requirements of Chapter 3 will not only allow the elimination of redundant text in other sections but may help to point out that as a general provision, clear floor spaces are permitted to overlap and serve multiple elements.

Allowing the clear floor spaces or clearances for the various elements to overlap helps allow designs to use less space if desired. Persons in wheelchairs need clear floor spaces, clearances, and turning spaces to enter and exit kitchens or toilet and bathing rooms/facilities, and to approach and utilize the facilities within them, but they typically utilize them one at a time. Allowing the clear floor spaces to overlap allows the elements to share the access or space that is needed for other elements while maintaining accessibility and reducing the space that is needed or dedicated to each individual element.

In most situations these changes will not result in any significant changes of application or enforcement when compared to the 2003 edition of the standard. One advantage of the changes may be that the requirements are clarified and will be applied more consistently. The inconsistent text caused people to wonder about how the provisions applied or forced them to look at several sections to decide how it should be read. For example, the Type A and Type B units contained this overlap allowance within their toilet and bathing facility requirements but they did not address this issue in their kitchen requirements. Therefore users of the previous standard had to decide if the overlap was permitted in kitchens even though it was not specifically addressed similar to the toilet and bathing facilities or to decide it was not prohibited by any text stating the overlap was not permitted.

One difference that may appear to be significant to some users is the inclusion of the door maneuvering clearance into the overlap section. As stated earlier, the people who use these clear spaces typically utilize them one at a time. Therefore prohibiting the maneuvering clearance at the entry door from overlapping any of the required clear floor spaces, clearances, or turning spaces within the room/facility does not appear to be warranted. However, because of the inclusion of door maneuvering clearance into this overlap section an additional modification was made to Section 404.2.3 so that maneuvering clearances at doors could not include the use of the knee and toe clearances that are allowed for fixtures. The change to Section 404.2.3 should assure that access to door hardware is not impeded and that users do not need to reach over an obstruction.

CHANGE TYPE: Modification

CHANGE SUMMARY: Provides clarification as to where the clear floor space must be located to be considered as an unobstructed reach to an element or control. Also limits the high reach for existing elements.

2009 STANDARD:

308.3 Side Reach.

308.3.1 Unobstructed. Where a clear floor space ~~complying with Section 305~~ allows a parallel approach to an element and the ~~side reach is unobstructed,~~ edge of the clear floor space is 10 inches (255 mm) maximum from the element, the high side reach shall be 48 inches (1220 mm) maximum and the low side reach shall be 15 inches (380 mm) minimum above the floor.

Exception: Existing elements that are not altered shall be permitted at 54 inches (1370 mm) maximum above the floor.

CHANGE SIGNIFICANCE: To provide an "unobstructed" reach to an element, the clear floor space must be located within 10 inches horizontally and the person must not need to reach over an object that would extend into the reach range height of 15 to 48 inches above the floor. This will help to distinguish when Section 308.3.1 for the "unobstructed" reach is used and when Section 308.3.2 for the "obstructed high reach" is applicable. By requiring the clear floor space to be located within the 10-inch horizontal distance and unobstructed, most users will be able to reach any element within the normal reach range limits.

The inclusion of this text will help to coordinate with the revised ADA and ABA Accessibility Guidelines. This revision was not intended to create any technical changes but was considered more as a clarification or interpretation of Figure 308.3.1 from the 2003 edition of the standard and Figure 6(b) from the original ADA Accessibility Guidelines (ADAAG). This will, however, make it obvious that in order to provide full access to an element a clear floor space must not be located more than 10 inches horizontally from the element.

Adding the text "that are not altered" into the exception helps to clarify and limit the application of the exception. If an existing element is altered it should comply with the 48-inch height limitation. Earlier editions of the A117.1 standard and the original ADAAG permitted a 54-inch high side reach. Because of this earlier allowance, numerous elements in existing buildings may be located above the 48-inch maximum height that is required by the base paragraph.

Existing elements may generally be left as they currently exist if they are not affected or changed during an alteration. The exception does, however, limit the height of these unaltered existing elements to the 54-inch height. If the existing element is above 54 inches then it would need to be lowered to the 48-inch maximum height. This revised exception will result in a slight difference between the A117.1 standard and the newly revised ADA and ABA Accessibility Guidelines (ADA and ABA AG). Under the new ADA and ABA AG there is no general exception that allows existing elements to be 54 inches high. In addition, under the federal law if it is technically infeasible to relocate an existing element there is no maximum height limit.

308.3.1
Unobstructed Side Reach Allowances

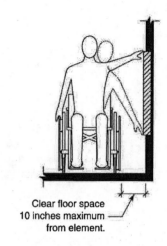

Clear floor space 10 inches maximum from element.

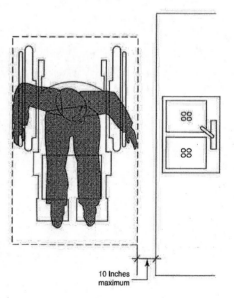

10 inches maximum

308.3.2

Side Reach Over an Obstruction, Appliance, or Counter

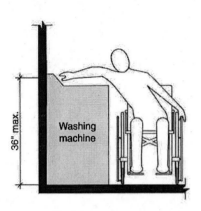

CHANGE TYPE: Modification

CHANGE SUMMARY: The new exception allows the height of the laundry equipment to be 36 inches instead of the normal height limit of 34 inches for reaching to the controls or into the appliance.

2009 STANDARD:

308.3 Side Reach.

308.3.2 Obstructed High Reach. Where a clear floor space <u>complying with Section 305</u> allows a parallel approach to an ~~object~~ <u>element</u> and the high side reach is over an obstruction, the height of the obstruction shall be 34 inches (865 mm) maximum <u>above the floor</u> and the depth of the obstruction shall be 24 inches (610 mm) maximum. The high side reach shall be 48 inches (1220 mm) maximum <u>above the floor</u> for a reach depth of 10 inches (255 mm) maximum. Where the reach depth exceeds 10 inches (255 mm), the high side reach shall be 46 inches (1170 mm) maximum <u>above the floor</u> for a reach depth of 24 inches (610 mm) maximum.

<u>**Exception:** At washing machines and clothes dryers, the height of the obstruction shall be permitted to be 36 inches (915 mm) maximum above the floor.</u>

CHANGE SIGNIFICANCE: The addition of the exception is the primary change within this section. The exception is included to conform to the revised ADA and ABA AG, which allow washers and dryers to exceed the maximum height requirements for obstructed reach. Otherwise, accessible machines would likely provide limited capacity as compared to inaccessible machines. There is slightly different wording between that of the ADA and ABA AG and the A117.1 to match the language in the charging paragraph.

The addition of the exception is also somewhat in recognition of the fact that the typical laundry equipment available on the market is 36 inches in height. It is this recognition of reality that was the basis for the concern that smaller equipment that could meet the 34-inch height limit would affect the capacity of the washer or dryer. With the increased popularity of front-loading machines, which often have the controls located at the front, there are more options available that can meet the general reach range requirements without the need to reach over an obstruction. Many of these front-loading machines provide a larger capacity and reduce water usage but may be initially more expensive than the traditional top-loading washer and matching dryer would be. Several additional revisions were made in Section 611 to address the front-loading laundry equipment.

The change from "object" to "element" replaces an undefined term with a term that is defined within the standard. This change should not create any change in application but is simply a clarification made for consistency with other parts of the standard. The addition of the phrase "complying with Section 305" was an editorial addition made for consistency and to provide the users with a link to the requirements that are applicable to that element.

CHANGE TYPE: Addition

CHANGE SUMMARY: Prohibits the use of Section 306 knee and toe clearances when establishing the clear maneuvering space required at a door. This should assure that access to door hardware is not impeded and that users do not need to reach over an obstruction to operate the hardware.

2009 STANDARD:

404.2.3 Maneuvering Clearances ~~at Doors~~. Minimum maneuvering clearances at doors shall comply with Section 404.2.3 and shall include the full clear opening width of the doorway. <u>Required door maneuvering clearances shall not include knee and toe clearance.</u>

CHANGE SIGNIFICANCE: This change was developed as a part of the A117 committee's discussion related to the overlap provisions of Section 301.2. Because clearances at fixtures can include knee and toe space beneath the fixture, there was concern that an element could be located within the door maneuvering space, which would require people to reach over an obstruction and therefore could significantly impede access to the door.

The previous edition of the standard did not address whether the knee and toe clearance provisions of Section 306 were allowed to be used when establishing the maneuvering clearances at doors. The lack of a specific statement led to inconsistent application of the provisions. While an element that required a full 25-inch-depth toe clearance would cause difficulty for many people to operate the door, the minor protrusion of items that did not require but simply provided a bit of clearance beneath it would not. For example, consider a small countertop extension in the area adjacent to a door or a handrail that runs along the wall. Both of these items may project over the floor area of the maneuvering space for the door but because the depth of the protrusion is limited and there is "toe clearance" beneath the element, access to the door is probably not affected. Personally, I used the 8-inch depth from the recessed door requirements to determine what obstructions were acceptable at this location while other users accepted the full 25-inch depth or no projections at all. Regardless of previous opinion, the standard now clearly addresses that knee and toe clearance is not permitted within the door maneuvering clearances.

404.2.3
Obstructions to the Maneuvering Clearances at Doors

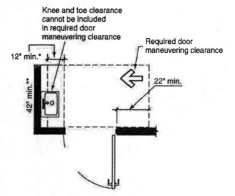

* If both closer and latch are provided
** 48" min. (1220) if both closer and latch provided

(e) Hinge Approach, Push Side

405.1, 405.2

Allowances for Ramps that are not Part of an Accessible Route

CHANGE TYPE: Modification

CHANGE SUMMARY: New exception is added to remind users that ramps that do not serve accessible elements are not required to comply with the ramp requirements of the standard.

2009 STANDARD:

405 Ramps

405.1 General. Ramps along accessible routes shall comply with Section 405.

Exception: In assembly areas, aisle ramps adjacent to seating and not serving elements required to be on an accessible route shall not be required to comply with Section 405.

405.2 Slope. Ramp runs shall have a running slope greater than 1:20 and not steeper than 1:12.

Exception: In existing buildings or facilities, ramps shall be permitted to have slopes steeper than 1:12 complying with Table 405.2 where such slopes are necessary due to space limitations.

TABLE 405.2 Allowable Ramp Dimensions for Construction in Existing Sites, Buildings, and Facilities

Slope[1]	Maximum Rise
Steeper than 1:10 but not steeper than 1:8	3 inches (75 mm)
Steeper than 1:12 but not steeper than 1:10	6 inches (150 mm)

[1] A slope steeper than 1:8 shall not be permitted.

CHANGE SIGNIFICANCE: The new exception will allow steeper slopes for ramps that serve seating and means of egress that are not required to be accessible. When an element is not required to be accessible or be on an accessible route, then it is not regulated by the standard but by some other document such as the International Building Code (IBC). Although the ramp requirements of the standard and the IBC will generally be coordinated, it is important to remember that not all ramps are regulated by the standard and therefore it is permissible in certain situations, such as those covered by the exception, that the ramps slopes are steeper than what is normally permitted for an accessible ramp.

The IBC permits ramps serving assembly seating and means of egress to have a slope up to 1 unit vertical for every 8 units horizontal (1:8) when the ramp is not a part of an accessible route. Additionally, this exception is in the ADA and ABA AG (Section 405.1) and will provide additional coordination with that document.

The added text within Section 405.2 serves as a reminder that by definition a ramp must have a slope steeper than 1:20. Therefore if the slope is 1 unit vertical in 20 units horizontal (1:20) or less, then it is not considered as being a ramp and is not regulated by Section 405. A sloped walking surface with a slope of 1:20 or less is regulated by the provisions of Section 403, while a "ramp" that has a slope steeper than 1:20 must conform to Section 405.

405.9

Edge Protection Along the Sides of Ramp Runs

CHANGE TYPE: Modification

CHANGE SUMMARY: The changes reformat the section for better clarity. The height of a curb serving as the edge protection is specified as 4 inches.

2009 STANDARD:

405.9 Edge Protection. Edge protection complying with Section 405.9.1 or 405.9.2 shall be provided on each side of ramp runs and at each side of ramp landings.

Exceptions:

1. ~~Ramps not required to have handrails where curb ramp flares complying with Section 406.3 are provided.~~

1. Edge protection shall not be required on ramps not required to have handrails and that have flared sides complying with Section 406.3.

2. Edge protection shall not be required on the sides of ramp landings serving an adjoining ramp run or stairway.

3. Edge protection shall not be required on the sides of ramp landings having a vertical drop-off of 1/2 inch (13 mm) maximum within 10 inches (255 mm) horizontally of the minimum landing area specified in Section 405.7.

4. Edge protection shall not be required on the sides of ramped aisles where the ramps provide access to the adjacent seats and aisle access ways.

405.9.1 Extended Floor Surface. The floor surface of the ramp run or ramp landing shall extend 12 inches (305 mm) minimum beyond the inside face of a railing complying with Section 505.

405.9.2 Curb or Barrier. A curb complying with Section 405.9.2.1 or a barrier complying with Section 405.9.2.2 shall be provided.

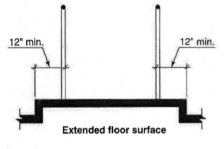

Extended floor surface

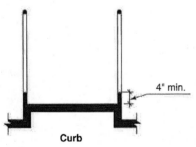

Curb

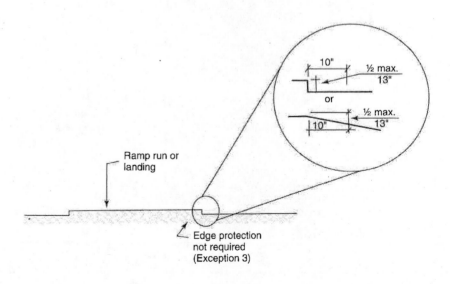

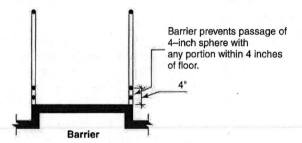

Barrier prevents passage of
4–inch sphere with
any portion within 4 inches
of floor.

4"

Barrier

405.9.2.1 Curb.   A curb shall be a minimum of 4 inches (100 mm) in height.

405.9.2.2 Barrier.   Barriers shall be constructed so that the barrier ~~that~~ prevents the passage of a 4-inch (100 mm) diameter sphere where any portion of the sphere is within 4 inches (100 mm) of the floor.

CHANGE SIGNIFICANCE: The intent of these sections is simply to prevent the small caster wheels of a wheelchair or the tips of crutches or other walking devices from going over the edge of a ramp. The provisions have been modified to help clearly show that there are three options for protecting the edge of the ramp. The options are the extended floor surface, a curb, or a barrier.

The format of the exceptions in Section 405.9 has been revised to help clarify that the exceptions are for the edge protection and to put the exceptions into complete sentences. Exception 1 is modified to coordinate with ADA and ABA AG Section 405.9 and is similar to that found in the 2003 A117.1 standard. This exception allows the elimination of edge protection when both of the conditions are met. This exception will only apply when the ramp has a rise of less than 6 inches, as stated in Section 405.8. The use of the flared sides for edge protection is acceptable for both curb ramps and other ramps provided the flares comply with the referenced requirements.

Exception 4 was added because ramps serving seating in assembly areas do not typically have drop-offs and would not have been covered by any of the other exceptions. The requirement for edge protection, if imposed at this location, would create a tripping hazard for people accessing the adjacent seating served by the ramp.

The primary technical change found within these sections is the fact that the standard will specify a minimum height for the curb. The previous standard was often viewed as not specifying the curb height. The committee discussion, however, stated that the performance statement about "*any* portion of the sphere" passing within 4 inches of the floor established the intended height. When the proposal to change this section was submitted it originally specified a 2-inch minimum curb height that matched earlier editions of the standard and to be consistent with the current ADA AG requirements. The 4-inch curb height is supposedly what is intended to be required by the new ADA and ABA AG, although the text of that document does not clearly include a minimum height requirement.

406.3

Curbs Along the Flared Sides of Curb Ramps

CHANGE TYPE: Addition

CHANGE SUMMARY: The new marking requirements help provide users with a visual clue to determine the location of the ramp, the curb, and where the curb flares are located. Where the adjacent curbs are painted, this section requires the flared portion of the curb also to be painted.

2009 STANDARD:

406 Curb Ramps

406.3 Sides of Curb Ramps. Where provided, curb ramp flares shall comply with Section 406.3.

406.3.1 Slope. Flares shall not be steeper than 1:10.

406.3.2 Marking. If curbs adjacent to the ramp flares are painted, the painted surface shall extend along the flared portion of the curb.

CHANGE SIGNIFICANCE: Curb ramps have undeniably helped provide accessible routes for those with different levels of physical disabilities, as well as other pedestrians, such as those pushing strollers or shopping carts. Unfortunately, the lack of visual markings on curb ramps can pose a fall hazard for persons with vision impairments.

The addition of 406.3.2 addresses the problem of pedestrians being exposed to visual miscues of the presence of drop-offs where one portion of the drop-off is painted and the adjacent portion of the drop-off, at the curb flare, is not. The absence of paint at the edge of the flare suggests to the user, approaching from the raised side and seeing the adjacent painted curb, that there is not a drop-off. The failure to perceive the drop-off at the flare can then lead to a misstep, often called an air step, and result in possible injury.

Where curbs are painted, many pedestrians use the painted curbs flanking a curb ramp as navigational cues to indicate the boundaries of a safe walkway. Where such visual clues do not exist or are not consistent it has resulted in stumbles, falls, and serious injuries as pedestrians mistakenly step off the flared side of the curb after initially perceiving it to be the

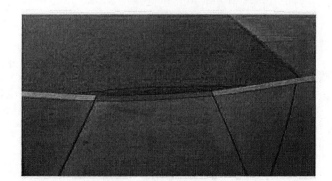

curb ramp itself. While the A117.1 standard previously provided extensive specifications for designing curb ramps, it failed to indicate how the flared side of curbs should be painted when the flanking curb is painted. This new text will not require that the curbs be painted, but where they are, it will specify how the curb at the flared sides of the curb ramp should be painted. The intent of the new wording is to indicate that it is the flared portion of the curb that is to be painted and that it is painted not only on the curb face but also on the top of the curb portion of the flare. It is not the intent to paint the flared sides, but rather just the curb portion of the flared sides. Painting the flared sides and/or curb ramp itself could make the ramp slippery when wet.

For additional information on this issue, users are directed to an article titled "Curb Ramps: Cross Slope Conspicuity and the Prevention of Air Steps," by Kenneth Nemire, HFE Consulting, Soquel, California. This article was from the Proceedings of the Human Factors and Ergonomics Society 50th Annual Meeting, 2006, and served as the basis for this text being added to the standard.

407.2.1.5

Call Button Signals at Elevator Landings

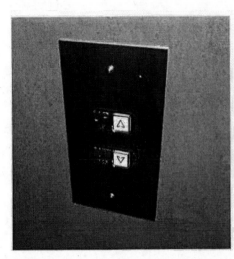

CHANGE TYPE: Addition

CHANGE SUMMARY: The new requirement provides two options so that a blind or visually impaired user will know that the activation of the elevator call button has been received.

2009 STANDARD:

407.2.1.5 Signals. Call buttons shall have visible signals to indicate when each call is registered and when each call is answered. <u>Call buttons shall provide an audible signal or mechanical motion of the button to indicate when each call is registered.</u>

Exceptions:

1. Destination-oriented elevators shall not be required to comply with Section 407.2.1.5, provided <u>a</u> visible <u>signal</u> and audible ~~signals~~ <u>tones and verbal announcements</u> complying with Section 407.2.1.7 are provided.
2. Existing elevators shall not be required to comply with Section 407.2.1.5.

CHANGE SIGNIFICANCE: Because the previously existing text of Section 407.2.1.5 provides a visual clue to users who are not blind or visually impaired, the new requirement is geared to the users who do not benefit from that visible signal. The intent of this added requirement is to provide improved access by providing either audible or tactile feedback of button activation. The net effect of this new text will be that heat-sensitive buttons or those that do not provide a discernible input would need to be capable of providing an audible indication that the button has been activated. This is somewhat similar to the requirements dealing with ATMs that preclude a touch-screen design where the user may not detect the activation of an input control.

While some of the committee discussion focused on the fact that the new text does not specify what the "mechanical motion" was required to be, the representatives from the elevator industry indicated that this type of system has been used effectively in Australia and that these types of products are readily available from the manufacturers. Although the committee agreed that the phrase "mechanical motion" would require some type of noticeable movement versus the use of a touch screen, there were no changes to the text to establish a specific requirement.

Perhaps the best solution when applying these requirements is that unless the call button responds and activates when there is a noticeable "mechanical motion," then the audible signal should be provided. A user who is not capable of noticing the visible signal would assume that if the call button did go through a noticeable mechanical motion that the call had been registered. Therefore care should be taken to avoid buttons that may provide a level of tactile feedback of the "mechanical motion" but that may not register the call if the button had not been pushed far enough to activate it. Until additional clarification of the requirements such as a specified level of performance for the mechanical motion is provided, designers and enforcers may want to assume that the call is registered by any movement of the button or that an audible signal is necessary.

The revisions within the exception should be viewed as an editorial clarification and not as a technical change or creating any additional requirement. This revision to "tones and verbal announcements" makes this exception, the provisions of Section 407.2.1.7 and also Exception 1 in Section 407.2.2.1 consistent with wording that has existed in Section 407.2.2.3, Exception 1. This will provide consistency for the destination-oriented elevator signal requirements.

407.4.6.1, 407.4.8

Elevator Car Control Location and Sequential Step Scanning

CHANGE TYPE: Modification

CHANGE SUMMARY: The revisions require that elevator buttons with floor designations be installed within the 48-inch reach range or that a system of sequential step scanning be installed. Previously, a 54-inch height was allowed for certain elevator control panels.

2009 STANDARD:

407.4.6 Elevator Car Controls. Where provided, elevator car controls shall comply with Sections 407.4.6 and 309.

Exception: In existing elevators, where a new car operating panel complying with Section 407.4.6 is provided, existing car operating panels shall not be required to comply with Section 407.4.6.

407.4.6.1 Location. Controls shall be located within one of the reach ranges specified in Section 308.

Exceptions:

1. Where the elevator panel ~~serves more than 16 openings and a parallel approach to the controls is provided, buttons with floor designations shall be permitted to be 54 inches (1370 mm) maximum above the floor~~ complies with Section 407.4.8 ~~Elevator Car Call Sequential Step Scanning~~.

2. In existing elevators, where a parallel approach is provided to the controls, car control buttons with floor designations shall be permitted to be located 54 inches (1370 mm) maximum above the floor. Where the panel is changed, it shall comply with Section 407.4.6.1.

407.4.8 Elevator Car Call Sequential Step Scanning. Elevator car call sequential step scanning shall be provided where car control buttons are provided more than 48 inches (1220 mm) above the floor~~, as permitted by Section 407.4.6.1, Exception #1~~. Floor selection shall be accomplished by applying momentary or constant pressure to the up or down scan button. The up scan button shall sequentially select floors above the current floor. The down scan button shall sequentially select floors below the current floor. When pressure is removed from the up or down scan button for more than 2 seconds, the last floor selected shall be registered as a car call. The up and down scan button shall be located adjacent to or immediately above the emergency control buttons.

CHANGE SIGNIFICANCE: The revisions will help make the requirements easier to understand and hopefully keep the requirements of Section 407.4.8 from being overlooked. Section 407.4.6.1 will require that elevator car control buttons be installed within the normal 15-inch minimum and 48-inch maximum reach range. Where the control buttons are not within this range, the standard will require that a sequential step scanning system be installed. This sequential step scanning system will allow users to select the floor level they wish to go to without having to reach for a button that is beyond the reach range.

In multistory buildings with numerous floor levels it is often difficult to get the elevator car control buttons arranged so that they all would fall within the required reach range. The sequential step scanning system permitted by Exception 1 will allow that device to be located within the required reach range and will mean that the car control panel itself can have buttons located outside of the reach range.

Previously the language of Section 407.4.6.1 Exception 1 allowed the control buttons to be located at a maximum 54-inch height when certain conditions existed. However, Section 407.4.8 imposed the requirement for sequential step scanning when the controls were beyond the 48-inch height. This requirement may have easily been overlooked if the user only looked at Exception 1 in 407.4.6.1. On the other hand, the language in Section 407.4.8 seemed to apply only to the situations allowed by Exception 1 in 407.4.6.1 instead of to all situations where the buttons were beyond the 48-inch reach range. By modifying both of the sections it is easier to understand the requirement and how the two sections work together.

This will improve access by assuring that the sequential step scanning system is installed whenever the car controls are beyond the reach range found in Section 308.

408.3.3

Door Location and Width for LULA Elevators

CHANGE TYPE: Modification

CHANGE SUMMARY: Provides additional options regarding the location and size of doors serving limited-use/limited-application (LULA) elevators. The new text within Section 408.3.3.2 allows doors on adjacent sides where the size of the elevator car is increased and a larger door is provided on the long side.

2009 STANDARD:

408.3.3 Door Location and Width. Car doors shall ~~provide a clear opening width of 32 inches (815 mm) minimum. Car doors shall positioned at a narrow end of the car~~ comply with Section 408.3.3.

408.3.3.1 Cars with Single Door or Doors on Opposite Ends. Car doors shall be positioned at the narrow end of cars with a single door and on cars with doors on opposite ends. Doors shall provide a clear opening width of 32 inches (815 mm) minimum.

408.3.3.2 Cars with Doors on Adjacent Sides. Car doors shall be permitted to be located on adjacent sides of cars that provide an 18 square foot (1.67 m²) platform. Doors located on the narrow end of cars shall provide a clear opening width of 36 inches (915 mm) minimum. Doors located on the long side shall provide a clear opening width of 42 inches (1065 mm) minimum and be located as far as practicable from the door on the narrow end.

Exception: Car doors that provide a clear opening width of 36 inches (915 mm) minimum shall be permitted to be located on adjacent sides of cars that provide a clear floor area of 51 inches (1295 mm) in width and 51 inches (1295 mm) in depth.

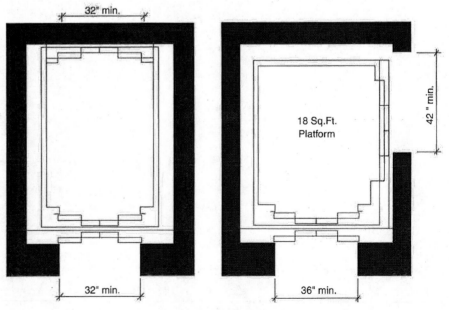

LULAs with Two Doors

CHANGE SIGNIFICANCE: The revisions provide more detailed requirements for LULAs that have doors on one end or have doors on opposite ends of the narrow side of the car. They also provide a new section that allows doors on adjacent sides if the size of the car is increased to 18 square feet in area. The provisions in 408.3.3.1 have essentially maintained the requirements from the 2003 edition of the standard but have added language to address the situation where doors are provided on opposite ends of the car. The previous standard would have accepted doors on opposite ends but the standard did not state whether this arrangement was permitted.

The requirements of Section 408.3.3.2 provide an alternative design option that goes beyond what was allowed by the previous exception to this section. This new section will require an increased elevator car floor area to provide adequate maneuvering clearances within the car where doors are installed on adjacent sides. The 18-square-foot car size coordinates with the maximum area that is permitted for a LULA elevator in Section 5.2.1.16.1 of ASME A17.1, but will also allow for adjustments in the width and depth of the car that are not allowed by the existing exception. This change allows for alternate designs where it is not possible to use a square car as required by the existing exception. For example, a car with a 48-inch width and a 54-inch depth would be permitted, while previously the only option was a 51-inch-by-51-inch car. It is important, however, to note that the new text will require a 36-inch clear opening on the narrow end and a 42-inch clear opening for the door on the long side of the elevator. The exception previously would have required only a 36-inch clear opening for the doors. It is this combination of increased car size, door width, and door location that will allow users sufficient space to maneuver through and access doors on adjacent sides.

408.4.1

Inside Dimensions of LULA Elevator Cars

CHANGE TYPE: Modification

CHANGE SUMMARY: Provides a minimum clear width requirement and a clear floor area requirement versus specifying both the minimum width and the minimum depth. Allows the elevator cars to be of any size within those limits.

2009 STANDARD:

408.4 Elevator Car Requirements. Elevator cars shall comply with Section 408.4.

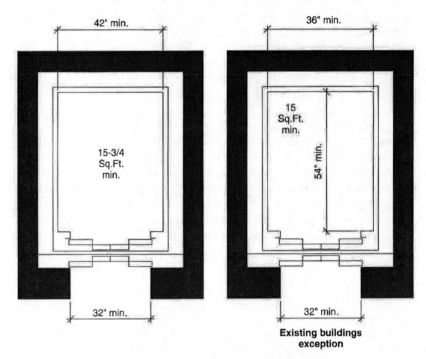

NOTE: Exception not applicable to cars with doors on adjacent sides.

408.4.1 Inside Dimensions of ~~Elevator Cars~~. Elevator cars shall provide a clear floor ~~area~~ width of 42 inches (1065 mm) minimum. ~~in width, and 54 inches (1370 mm) minimum in depth~~ The clear floor area shall not be less than 15-3/4 square feet (1.46 m^2).

Exceptions:

1. ~~Cars that provide a 51 inches (1295 mm) minimum clear floor width shall be permitted to provide 51 inches (1295 mm) minimum clear floor depth.~~

2. ~~For~~ installations in existing buildings, elevator cars that provide a clear floor area of 15 square feet (1.4 m^2) minimum, and provide a clear inside dimension of 36 inches (915 mm) minimum in width and 54 inches (1370 mm) minimum in depth, shall be permitted. This exception shall not apply to cars with doors on adjacent sides.

CHANGE SIGNIFICANCE: The change to the base paragraph of this section will allow more options for elevator car sizes and therefore may provide additional solutions and improved access for existing buildings that install a LULA. The revision eliminates the minimum depth requirement and replaces it with a minimum area requirement. When combined with the 18-square-foot maximum area requirement found in ASME A17.1 for these elevators, it allows car sizes of any dimension so long as the minimum clear width requirement and the minimum and maximum clear floor area criteria are met. It should be noted that although the minimum depth is not specified in this section, a 48-inch minimum would be required in order to satisfy the clear floor space requirements of Section 305.3.

The 15-3/4-square-foot area in the new text is equivalent to what the previous edition required with the minimum 42-inch width and 54-inch depth. However, the previous language would not have permitted the installation of a car size of 50 inches in width and 52 inches in depth even though the dimensions would have been usable for access and within the maximum allowed for these elevators. This 50-inch-by-52-inch car size would not have even been permitted within an existing building even though the floor area is larger than that required previously by Exception 2. When looked at in conjunction with the changes covered earlier in Section 408.3.3, establishing the parameters of minimum width along with minimum and maximum total floor area is a better, more inclusive approach to regulating the car size.

The exceptions have been modified because of this change and the requirements added into Section 408.3.3. Exception 1 was eliminated because the 51-inch-by-51-inch size that was previously permitted by this exception, and is still permitted by the exception in Section 408.3.3.2, would be allowed by the new language in the base paragraph. (For those people who like to be 100% accurate: The car size is considered to be 18 square feet in area even though it is actually 18.06 square feet.) The language from the previous Exception 2 has been changed so that it does not apply to cars with doors on adjacent sides. This coordinates with the revision discussed earlier in this book with Section 408.3.3 and the fact that it is difficult to maneuver the turn through doors on adjacent sides when the car size is less than the maximum allowed. Therefore any elevator installed in an existing building using the exception will need to have the door or doors positioned on the narrow ends of the car as required by Section 408.3.3.1.

409.3.1

Power Operation of Doors and Gates for Private Residence Elevators

CHANGE TYPE: Modification

CHANGE SUMMARY: The revision limits the allowance for a manually operated door to situations where the user would enter the car and need to open the door on the opposite side by only using a forward approach.

2009 STANDARD:

409.3 Doors and Gates. Elevator car and hoistway doors and gates shall comply with Sections 409.3 and 404.

Exception: The maneuvering clearances required by Section 404.2.3 shall not apply for approaches to the push side of swinging doors.

409.3.1 Power Operation. Elevator car doors and gates shall be power operated and shall comply with ANSI/BHMA A156.19 listed in Section 105.2.3. Elevator cars with a single opening shall have low energy power operated hoistway doors and gates.

Exception: ~~For elevators with a car that has more than one opening, the hoist way doors and gates shall be permitted to be of the manual open, self close type.~~ Hoistway doors or gates shall be permitted to be of the self-closing, manual type, where that door or gate provides access to a narrow end of the car that serves only one landing.

CHANGE SIGNIFICANCE: This revision will provide a clearer and more practical limit regarding when a manual door or gate is permitted for the private residence elevator. The wording of the exception allows a manual

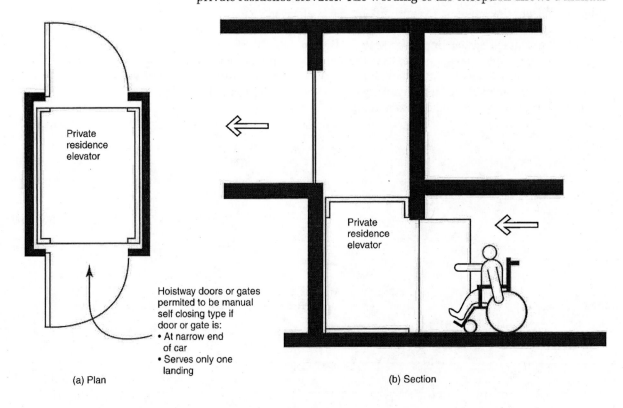

Hoistway doors or gates permited to be manual self closing type if door or gate is:
- At narrow end of car
- Serves only one landing

(a) Plan

Private residence elevator

Private residence elevator

(b) Section

door or gate where the elevator car has the doors or gates at the narrow ends of the car and the opening serves only one landing. The intent of the text limiting the door to serving only one landing is for the situation where you entered using a forward approach, moved up or down, and then to exit you had to move backwards (i.e., the door serves two landings); in that situation the door would have to be automatic. If you enter forward, and then exit forward, each door is serving only one landing and the exception is applicable. By limiting the doors to the narrow end and allowing them to serve only one landing, it ensures that the user would be able to enter the elevator car and then exit at the next level by continuing in a forward direction. With this arrangement and limitation the user would not need to maneuver, turn around, or reach behind him- or herself to operate the door at another level.

The previous language would have allowed for a situation where the door or gate at one of the ends could have served two or more landings. Such situations would have resulted in the user having to maneuver while in the elevator or to reach behind him- or herself to operate the door that he or she had used to enter.

Where the doors or gates can be operated while continuing in the same direction, it is reasonable to allow for their manual operation. Where users would need to maneuver, turn around, or be expected to reach behind them to operate the doors, the exception would not be applicable and power operation in compliance with the BHMA A156.19 standard is required. This change will correspond to a similar revision that was made to Section 410.2.1 Exception 1 and applicable to platform lifts.

409.3.3

Door or Gate Location and Width for Private Residence Elevators

CHANGE TYPE: Modification

CHANGE SUMMARY: Provides additional design options for private residence elevators by allowing doors to be located at the side of the elevator car instead of limiting them to the narrow end. Provisions also provide specific clearances for the doors.

2009 STANDARD:

409.3.3 Door or Gate Location <u>and Width.</u> Car gates or doors ~~shall be~~ positioned at a narrow end of the clear floor area required by Section 409.4.1 <u>shall provide a clear opening width of 32 inches (815 mm) minimum. Car gates or doors positioned on adjacent sides shall provide a clear opening width of 42 inches (1065 mm) minimum.</u>

CHANGE SIGNIFICANCE: Private residence elevators are becoming more common in townhomes and other applications that may fall under the requirements for Type B units. This revision provides additional flexibility when designing a unit with an elevator by allowing options for the locations of the doors. Allowing the possibility of a door on the side of the elevator car will allow more residences to accommodate the installation of these elevators.

Because the minimum inside dimensions are established by Section 409.4.1 and the maximum size of these elevators is limited to 15 square feet by the ASME A17.1 standard, the design options and the possibility for maneuvering space are greatly restricted.

Where the door is located on the long side of the elevator, the clear width of the door must be increased and all the doors must be power assisted or a low-energy power-operated door in accordance with Section 409.3.1. This increased clearance and power operation will help allow for maneuvering onto and off of the elevator.

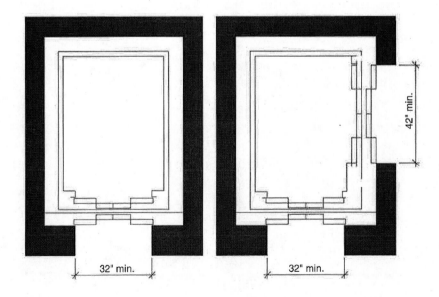

CHANGE TYPE: Modification

CHANGE SUMMARY: These changes address maneuvering and functioning within lifts that have doors on adjacent sides. They also address lift doors or gates with ramps.

410.2 , 410.5

Lift Entry and Clear Floor Space for Platform Lifts

2009 STANDARD:

410.2 Lift Entry.   Lifts with doors or gates shall comply with Section 410.2.1. Lifts with ramps shall comply with Section 410.2.2.

410.2.1 Doors and Gates.   Doors and gates shall be low-energy power-operated doors or gates complying with Section 404.3. Doors shall remain open for 20 seconds minimum. <u>On lifts with one door or with doors on opposite ends, the</u> end door clear opening width shall be 32 inches (815 mm) minimum. <u>On lifts with one door on a narrow end and one door on a long side, the end door clear opening width shall be 36 inches (915 mm) minimum.</u> Side door clear opening width shall be 42 inches (1065 mm) minimum. <u>Where a door is provided on a long side and on a narrow end of a lift, the side door shall be located with either the strike side or the hinge side in the corner furthest from the door on the narrow end.</u>

Exceptions:

1. ~~Lifts serving two landings maximum and having doors or gates on opposite sides shall be permitted to have self-closing manual doors or gates.~~ <u>Doors or gates shall be permitted to be of the self-closing, manual type, where that door or gate provides access to a narrow end of the platform that serves only one landing. This exception shall not apply to doors or gates with ramps.</u>

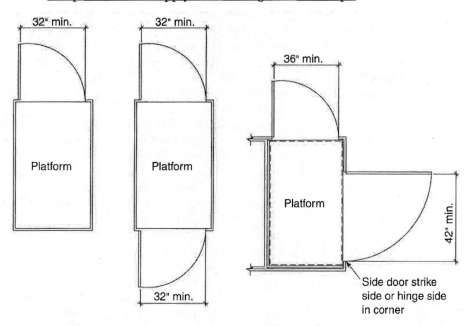

(a) Platform lift with door at one end or opposite ends

(b) Platform with doors on adjacent sides

410.2, 410.5 continues

410.2, 410.5 continued

2. Lifts serving two landings maximum and having doors or gates on adjacent sides shall be permitted to have self-closing manual doors or gates provided that the side door or gate is located with the strike side furthest from the end door. This exception shall not apply to doors or gates with ramps.

410.2.2 Ramps. End ramps shall be 32 inches (815 mm) minimum in width. Side ramps shall be 42 inches (1065 mm) minimum in width. Ramp widths shall not be less than the platform opening they serve.

410.5 Clear Floor Space. Clear floor space of platform lifts shall comply with Section 305 410.5.

410.5.1 Lifts with Single Door or Doors on Opposite Ends. Platform lifts with a single door or with doors on opposite ends shall provide a clear floor width of 36 inches (915 mm) minimum and a clear floor depth of 48 inches (1220 mm) minimum.

410.5.2 Lifts with Doors on Adjacent Sides. Platform lifts with doors on adjacent sides shall provide a clear floor width of 42 inches (1065 mm) minimum and a clear floor depth of 60 inches (1525 mm) minimum.

Exception: In existing buildings, platform lifts with doors on adjacent sides shall be permitted to provide a clear floor width of 36 inches (915 mm) and a clear floor depth of 60 inches (1525 mm).

CHANGE SIGNIFICANCE: These changes address maneuvering and functioning within lifts that have doors on adjacent sides. They also address lift doors or gates with approach ramps.

The text of Section 410.2.1 is revised so that a lift with a side door and a door on the narrow end will require an increased clear width for the end door. While doors on the narrow end are generally required to provide a 32-inch clear width, those with adjacent side doors will require a 36-inch clear width. This extra width will provide additional maneuvering space for the 90-degree turn that is needed to get through the side door. Conceptually this is similar to the T-shaped turning space in Section 304 and makes maneuvering through the door easier when the user must move to or from the side to use the door.

The text of this base section was also revised to require that either the strike side or hinge side edge of the side door be located in the corner of the lift as far from the end door as possible. The side door is required to be located as far as possible from the end door to provide the maximum maneuvering space on the platform. Because the side doors are generally required to be low-energy power-operated doors, the direction of the door swing is not a major concern. However, Exception 2 will require that the strike side of the side door be located in the far corner when following those requirements and using a manual door or gate. Lift platforms are not large enough to provide the required maneuvering clearance to a door approached from the latch side. Therefore, Exception 2 requires that manual side doors be located with a hinge side approach while on the lift.

Exception 1 will accept manual doors at the ends where the user is not required to make a turn going out of the door. The operation of doors at the ends is similar to normal door operation. The intent of the text limiting the door to serving only one landing is that the user would be approaching the door in a forward direction. If the door served more than one landing it could result in locations where the door could be at your back and would be required to be automatic. For the situation where you entered forward, moved up or down, and then to exit you had to move backwards (i.e., the door serves two landings), then the door would have to be automatic. If you entered forward, could travel to only one other landing, and then exited forward, each door is serving only one landing and Exception 1 is applicable.

Section 410.2.2 was revised to tie the width of the approach ramp to the size of the platform opening versus simply establishing a minimum width as previously done. As previously written, an approach ramp could have been narrower than the door or gate that served it. An approach ramp that is narrower than the opening would clearly create an unsafe condition.

The changes to Section 410.5 provide more detailed criteria for lifts with doors on one or opposite sides and for lifts with doors on adjacent sides. The previous reference to the clear floor space requirements of Section 305 could not adequately address the need for maneuvering onto and off of the lift. The width of 36 inches found in Section 410.5.1 provides consistency with the alcove and accessible route requirements found in Chapters 3 and 4 of the standard. When entering and exiting, the size must be 36 inches for an alcove, and when rolling through the accessible route must also be 36 inches wide if it is more than 24 inches in depth.

Where doors on the lifts are on adjacent sides the new text in Section 410.5.2 requires the platform size to be increased in order to provide adequate maneuvering clearance on the platform. This section is conceptually similar to the revisions found in the LULA elevator section but it does require a larger size platform with a 60-inch minimum depth to make it work. The 60-inch depth was selected to coordinate with the alcove requirements of Section 305.7. Where the lift is being installed in an existing building the exception will help address situations where it may not be possible to accommodate a normally sized lift. The allowance for a smaller lift does, however, still retain the 60-inch length requirement to provide adequate space when entering or exiting the lift from the side door. This extra length allows the user to complete a 90-degree turn onto the lift and remains coordinated with the alcove requirements. Users should remember that the provisions of Section 410.5.1 will provide a smaller lift that is also permitted to be used for existing buildings if the 36-inch-by-60-inch platform allowed by the exception will not work.

602.2

Clear Floor Space at Drinking Fountains

CHANGE TYPE: Deletion

CHANGE SUMMARY: Two exceptions that previously allowed for a parallel approach to existing drinking fountains have been deleted.

2009 STANDARD:

602 Drinking Fountains

602.1 General. Accessible drinking fountains shall comply with Sections 602 and 307.

602.2 Clear Floor Space. A clear floor space complying with Section 305, positioned for a forward approach to the drinking fountain, shall be provided. Knee and toe space complying with Section 306 shall be provided. The clear floor space shall be centered on the drinking fountain.

Exceptions:

1. Drinking fountains for standing persons.
2. Drinking fountains primarily for children's use shall be permitted where the spout <u>outlet</u> is 30 inches (760 mm) maximum above the floor, ~~and~~ a parallel approach complying with Section 305 <u>is provided, and the clear floor space is</u> centered on the drinking fountain~~, is provided.~~
3. ~~In existing buildings, existing drinking fountains providing a parallel approach complying with Section 305, centered on the drinking fountain, shall be permitted.~~
4. ~~Where specifically permitted by the administrative authority, a parallel approach complying with Section 305, centered on the drinking fountain, shall be permitted for drinking fountains that replace existing drinking fountains with a parallel approach.~~

CHANGE SIGNIFICANCE: Exceptions 3 and 4 were deleted from the standard in an attempt to harmonize with the revised ADA and ABA AG, which does not allow a parallel approach to drinking fountains in existing facilities. If these exceptions were not removed, the A117.1 standard would have been less stringent than the federal law and these exceptions would have led to potential problems for people who relied upon the standard when evaluating compliance with accessibility requirements.

These two previous exceptions provided blanket exemption for both new replacement and existing drinking fountains that are not necessarily based on technical infeasibility. Both of these exceptions potentially conflict with the requirements of Section 202.4 (Alterations Affecting Primary Function Areas) of the revised ADA and ABA AG, as well as Sections 605 and 605.2 (Alterations Affecting an Area Containing a Primary Function) of the International Existing Building Code (IEBC). Each of these two documents already provides exceptions for existing drinking fountains. However, instead of providing a blanket exemption, these two documents require that the design be looked at based on the issues of whether it is technically infeasible and at the level of work that is being done.

CHANGE TYPE: Modification and clarification

CHANGE SUMMARY: New text ensures that the required turning space is not located within a toilet compartment and potentially unavailable for use. The provisions dealing with the overlap of clear floor spaces and clearances has been moved to Chapter 3.

603.2

Clearances in Toilet and Bathing Rooms

2009 STANDARD:

603 Toilet and Bathing Rooms

603.1 General.   Accessible toilet and bathing rooms shall comply with Section 603.

603.2 Clearances.

603.2.1 Turning Space.   A turning space complying with Section 304 shall be provided within the room. <u>The required turning space shall not be provided within a toilet compartment.</u>

603.2.2 Overlap.   ~~Clear floor spaces, clearances at fixtures, and turning spaces shall be permitted to overlap.~~

603.2.2 Door Swing.   Doors shall not swing into the clear floor space or clearance for any fixture.

Exceptions:

1. Doors to a toilet ~~and~~ <u>or</u> bathing room for a single occupant, accessed only through a private office and not for common use or public use, shall be permitted to swing into the clear floor space, provided the swing of the door can be reversed to ~~meet~~ <u>comply with</u> Section 603.2.2.

2. Where the room is for individual use and a clear floor space complying with Section 305.3 is provided within the room beyond the arc of the door swing<u>, the door shall not be required to comply with Section 603.2.2.</u>

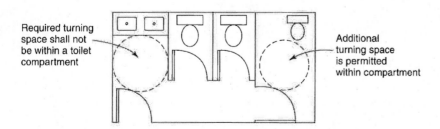

Required turning space shall not be within a toilet compartment

Additional turning space is permitted within compartment

603.2 continues

603.2 continued

CHANGE SIGNIFICANCE: The previous language did not restrict the location of the required turning space. It would therefore have been permissible to place the required turning space within an enlarged toilet compartment even though that would clearly affect the overall use of the space and the other fixtures. The intent of the requirement is that the turning space needs to be provided within the room but clear of the compartment in order to provide a functional and usable room. It is important for accessible lavatories, other fixtures, and even the door from the room to be available to users of the facility without requiring them to access the toilet compartment in order for them to turn and be able to use them.

It should be noted that the text would not prohibit a toilet compartment from containing a turning space; it only prohibits the required turning space for the room from being located within the compartment. Therefore if a turning space is provided in the main portion of the room, perhaps near the lavatories, then an additional turning space could be provided within an individual toilet compartment.

The overlap requirement that was previously in Section 603.2.2 has simply been relocated to the general requirements of Chapter 3. This change was discussed earlier in this document with the changes in Section 301.2. Providing a single location for this requirement will make it more apparent and eliminate the confusion that was caused when this text was included in some locations and not within others.

The change to Exception 1 in Section 603.2.2 and a corresponding change in Section 606.3 help to clarify the application of the exception. By using the word "and," the previous language seemed to imply that the room needed to contain both bathing *and* toilet facilities before it was qualified to use the exception. The intent of the previous edition and the purpose for this change is to clarify that *either* a toilet room *or* a bathing room for a single occupant could qualify for the application of the exception. One additional benefit of this clarification is that it coordinates the standard with the provisions of the ADA and ABA AG.

603.3

Mirrors in Toilet and Bathing Rooms

CHANGE TYPE: Modification

CHANGE SUMMARY: This revision provides better clarity regarding where an accessible mirror is required to be installed within toilet or bathing rooms. An exception will allow the use of a wall-mounted mirror as an alternative in public facilities, while within Accessible and Type A units mirrors installed above the accessible lavatory must be accessible.

2009 STANDARD:

603.3 Mirrors. Where mirrors are located above lavatories, sinks or counters a mirror shall be located over the accessible lavatory and shall be mounted with the bottom edge of the reflecting surface 40 inches (1015 mm) maximum above the floor. Where mirrors are located above counters that do not contain lavatories, a mirror shall be mounted with the bottom edge of the reflecting surface 40 inches (1015 mm) maximum above the floor.

Exception: Other than within Accessible dwelling or sleeping units, mirrors not located above lavatories, sinks or counters shall be are not required over the lavatories or counters if a mirror is located within the same toilet or bathing room and mounted with the bottom edge of the reflecting surface 35 inches (890 mm) maximum above the floor.

1002 Accessible Units

1002.11.2.2 Mirrors. Mirrors above accessible lavatories shall have the bottom edge of the reflecting surface 40 inches (1015 mm) maximum above the floor.

603.3 continues

1003.11.2.3 Mirrors. Mirrors above <u>accessible</u> lavatories shall have the bottom edge of the reflecting surface 40 inches (1015 mm) maximum above the floor.

CHANGE SIGNIFICANCE: The reformatting and revisions made to these sections should provide better clarity regarding the requirement for a mirror to be accessible and also what mirrors are not regulated.

In Section 603.3, the first sentence would put an accessible mirror over the accessible lavatory when mirrors were provided over non-accessible lavatories. This first sentence was also revised to eliminate "sinks or counters." Section 603 deals with toilet and bathing rooms; therefore it is not likely that a "sink" would be located in such spaces. The assumption is that lavatories are used in these locations and that "sinks" would be found in kitchens and perhaps janitorial closets. Counters are now addressed by the new text in the base paragraph. Since there is not a requirement for a portion of a counter to be accessible, where a mirror is provided over a counter the entire mirror will generally be installed at the 40-inch maximum height and be accessible. The portion that is accessible would generally be at the designer's discretion but for aesthetic reasons this typically would be the entire mirror. While not specifically stated, if the counter does have a portion set at an accessible height (i.e., work surface), that portion should include the accessible mirror.

As an exception to the accessible mirror being located over a lavatory or counter, a wall mirror would be an alternative due to the creation of a new exception. This new exception was created from both new and existing text. The added language makes sure the accessible mirror is in the bathing/grooming area. Though not specifically stated, it should be assumed that the intent is that the mirror is located in the general room area and should not be located within a toilet compartment or other isolated location. The use of the exception is acceptable in public bathrooms but not within accessible dwelling or sleeping units.

The text in Section 1002 dealing with the mirrors in accessible units coordinates with the exception in Section 603.3 and would exclude the option of using a wall mirror instead of a mirror over the accessible lavatory. Notice that these sections do not require the installation of a mirror but that they simply regulate the mirror when it is installed above the accessible lavatory. The intent is that within an accessible unit the required accessible toilet and bathing facility would have an accessible mirror if the mirror is mounted over a lavatory as is typical for dwelling units. Where there are multiple lavatories installed the standard would only regulate the mirror over the one required lavatory.

The requirement for the Type A unit is similar in intent in that within a unit with two lavatories only one of them is required to be accessible. It is only the mirror that is above the accessible lavatory that is regulated. The Type B units did not get modified by this change because the Fair Housing Act does not address mirrors in the bathrooms.

CHANGE TYPE: Addition

CHANGE SUMMARY: New section requires that diaper changing tables comply with both the operable parts and work surface requirements of the standard.

603.5
Diaper Changing Tables

2009 STANDARD:

603.5 Diaper Changing Tables.   Diaper changing tables shall comply with Sections 309 and 902.

CHANGE SIGNIFICANCE: This new text addresses an element that is commonly found within toilet and bathing rooms but was not specifically regulated. It would have generally been considered a work surface by the previous editions of the standard. This section provides the technical requirements for the changing tables and should not be viewed as requiring the tables to be installed. Where changing tables are installed or are required to be installed by a scoping document, the standard would regulate them as it would any other work surface or operable part. For example, Section 1109.2 of the IBC requires that "at least one of each type of fixture, element, control or dispenser in each accessible toilet room and bathing room shall be accessible." Therefore where a single diaper changing table is installed in the toilet or bathing room it must be installed so that it is accessible. It is important to realize that this requirement regulates not only built-in changing tables but also any kit or component-type table that is simply mounted on the wall after the completion of the construction.

By referencing Sections 309 and 902 this section requires that the changing table is provided with a forward-approach clear floor space, is located at the proper height, and that any latch or other operable part is capable of meeting the operation requirements of Section 309.4 and located within the proper reach range. Where the changing table is a

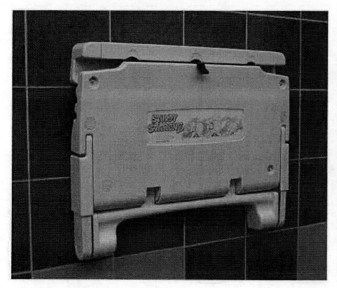

603.5 continues

603.5 continued folding type, this means that the forward-approach clear floor space (Section 902.2) must be available both when the table is in the stored position and also when it is down in the usable position. This may affect the mounting height since knee and toe clearance would be required beneath the table and yet the top "work surface" must not be higher than the 34-inch maximum height (Section 902.4). The latch to release the unit from its stored position would need to fall within the required reach ranges (Section 309.3).

CHANGE TYPE: Modification

603.6

Operable Parts of Towel Dispensers and Hand Dryers

CHANGE SUMMARY: Section 606.7 from the previous standard has been moved to Section 603.6. Moving these requirements to this section will make it more apparent that the towel dispensers and hand dryers need to be located so that the reach depth over an obstruction is limited.

2009 STANDARD:

603.6 ~~606.7~~ Operable Parts.   Operable parts on towel dispensers and hand dryers <u>serving accessible lavatories</u> shall comply with Table ~~606.7~~ <u>603.6</u>.

TABLE <u>603.6</u> ~~606.7~~ Maximum Reach Depth and Height

Maximum Reach Depth	0.5 inch (13 mm)	2 inches (50 mm)	5 inches (125 mm)	6 inches (150 mm)	9 inches (230 mm)	11 inches (280 mm)
Maximum Reach Height	48 inches (1220 mm)	46 inches (1170 mm)	42 inches (1065 mm)	40 inches (1015 mm)	36 inches (915 mm)	34 inches (865 mm)

CHANGE SIGNIFICANCE: Technically this is not a significant change as far as the A117.1 standard is concerned. The requirements of this section will be applicable exactly as they were in the previous edition of the standard. From the A117.1 viewpoint this section has simply been relocated to the general bathroom requirements of Section 603 and is no longer buried within the lavatory requirements of Section 606.

The significance of this change is the relocation of the provisions within the standard and the 2009 IBC including scoping provisions related to "enhanced reach range" lavatories. These two factors make it more apparent that this requirement exists, and therefore compliance with and enforcement of the provision will dramatically change. Though not technically tied to the enhanced reach range lavatories (A117.1 Section 606.5), this provision was often viewed as having that connection or simply overlooked due to its location at the end of the lavatory section. The addition

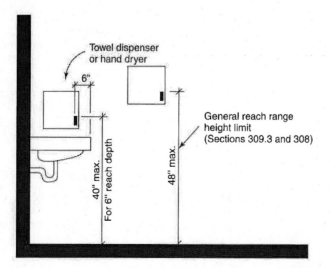

603.6 continues

603.6 continued

of scoping requirements within the IBC for the enhanced reach range lavatory will lead people to look at the standard a bit closer, and these requirements for the towel dispensers and hand dryers will be more apparent.

Users will need to determine how these reach requirements for the towel dispenser and hand dryer should be applied where there are multiple lavatories within a toilet or bathing room. In general, if the towel dispenser is located with a specific lavatory it would probably be best to make the towel dispenser that is with the accessible lavatory be the one that is required to comply with these reach ranges. In a larger toilet or bathing room where the IBC would require an "enhanced reach range" lavatory (rooms with six or more lavatories within them), it would probably be appropriate to make these limited reach towel dispensers and hand dryers serve that lavatory. Although there is nothing in the standard or the IBC that requires this relationship between the accessible lavatory and the regulated towel dispenser, it would seem appropriate to apply the requirements in such a manner. Realize, however, that the towel dispensers are often not located with or over a lavatory but are instead located on an end wall near the door or on the back wall behind the user when the user is facing the lavatory. This type of arrangement where the dispenser or hand dryer is not associated with a particular lavatory is perfectly acceptable. In such arrangements it may be easier to ensure that one of the towel dispensers or hand dryers does meet the limited reach requirements of Table 603.6.

These requirements were originally developed and included in the A117.1 standard based on the efforts of the A117 committee member organization Little People of America (LPA). LPA represents people of short stature, including those with various forms of dwarfism. Due to the limited reach of people of short stature as well as those people in wheelchairs or children, it is important to locate these fixtures so that the reach depth over an obstruction is limited. While the relocation of these provisions is not a significant technical change, any improved application or compliance with the requirements will provide a significant improvement in accessibility for these users.

CHANGE TYPE: Modification

CHANGE SUMMARY: Blocking requirements for grab bars that are applicable to the various types of dwelling units have been relocated to Chapter 10.

604.5, 607.4

Grab Bar Blocking and Swing-up Grab Bar Requirements

2009 STANDARD:

604.5 Grab Bars. Grab bars for water closets shall comply with Section 609 and shall be provided in accordance with Sections 604.5.1 and 604.5.2. Grab bars shall be provided on the rear wall and on the side wall closest to the water closet.

Exceptions:

1. Grab bars are not required to be installed in a toilet room for a single occupant, accessed only through a private office and not for common use or public use, provided reinforcement has been installed in walls and located so as to permit the installation of grab bars complying with Section 604.5.

2. In detention or correction facilities, grab bars are not required to be installed in housing or holding cells or rooms that are specially designed without protrusions for purposes of suicide prevention.

3. ~~In Type A units, grab bars are not required to be installed where reinforcement complying with Section 1003.11.4 is installed for the future installation of grab bars.~~

4. ~~In Type B units located in institutional facilities and assisted living facilities, two swing-up grab bars shall be permitted to be installed in lieu of the rear wall and side wall grab bars. Swing-up grab bars shall comply with Sections 604.5.3 and 609.~~

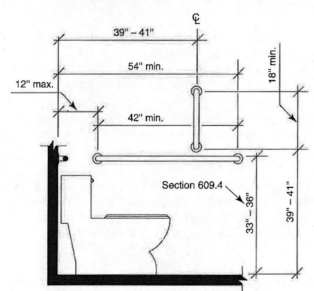

Note: Dimensions shown are for adult water closets.

604.5, 607.4 continues

604.5, 607.4 continued

5. ~~In a Type B unit, where fixtures are located on both sides of the water closet, a swing-up grab bar complying with Sections 604.5.3 and 609 shall be permitted. The swing-up grab bar shall be installed on the side of the water closet with the 18 inch (455 mm) clearance required by Section 1004.11.3.1.2.~~

604.5.1 Fixed Side Wall Grab Bars. Fixed side wall grab bars shall be 42 inches (1065 mm) minimum in length, located 12 inches (305 mm) maximum from the rear wall and extending 54 inches (1370 mm) minimum from the rear wall. In addition, a vertical grab bar 18 inches (455 mm) minimum in length shall be mounted with the bottom of the bar located ~~between~~ 39 inches (990 mm) <u>minimum</u> and 41 inches (1040 mm) <u>maximum</u> above the floor, and with the center line of the bar located ~~between~~ 39 inches (990 mm) <u>minimum</u> and 41 inches (1040 mm) <u>maximum</u> from the rear wall.

Exceptions: <u>The vertical grab bar at water closets primarily for children's use shall comply with Section 609.4.2.</u>

1. ~~In Type A and Type B units, the vertical grab bar component is not required.~~
2. ~~In a Type B unit, when a side wall is not available for a 42-inch (1065 mm) grab bar, the sidewall grab bar shall be permitted to be 18 inches (455 mm) minimum in length, located 12 inches (305 mm) maximum from the rear wall and extending 30 inches (760 mm) minimum from the rear wall.~~

604.5.2 Rear Wall Grab Bars. The rear wall grab bar shall be 36 inches (915 mm) minimum in length, and extend from the centerline of the water closet 12 inches (305 mm) minimum on the side closest to the wall, and 24 inches (610 mm) minimum on the transfer side.

Exceptions:

1. The rear grab bar shall be permitted to be 24 inches (610 mm) minimum in length, centered on the water closet, where wall space does not permit a grab bar 36 inches (915 mm) minimum in length due to the location of a recessed fixture adjacent to the water closet.
2. ~~In a Type A or Type B unit, the rear grab bar shall be permitted to be 24 inches (610 mm) minimum in length, centered on the water closet, where wall space does not permit a grab bar 36 inches (915 mm) minimum in length.~~
3. ~~3.~~ <u>2.</u> Where an administrative authority requires flush controls for flush valves to be located in a position that conflicts with the location of the rear grab bar, that grab bar shall be permitted to be split or shifted to the open side of the toilet area.

604.5.3 Swing-up Grab Bars. ~~Where swing-up grab bars are installed, a clearance of 18 inches (455 mm) minimum from the centerline of the water closet to any side wall or obstruction shall be provided. A swing-up~~

~~grab bar shall be installed with the centerline of the grab bar 1 5¾ inches (400 mm) from the centerline of the water closet. Swing-up grab bars shall be 28 inches (710 mm) minimum in length, measured from the wall to the end of the horizontal portion of the grab bar.~~

607.4 Grab Bars. Grab bars shall comply with Section 609 and shall be provided in accordance with Sections 607.4.1 or 607.4.2.

Exceptions:

~~1.~~ Grab bars shall not be required to be installed in a bathing facility for a single occupant accessed only through a private office and not for common use or public use, provided reinforcement has been installed in walls and located so as to permit the installation of grab bars complying with Section 607.4.

~~2. In Type A units, grab bars are not required to be installed where reinforcement complying with Section 1003.11.4 is installed for the future installation of grab bars.~~

CHANGE SIGNIFICANCE: By deleting these requirements from Chapter 6 and relocating them to Chapter 10, it will allow the provisions to be with the type of dwelling unit that each of the provisions applies to. This allows the requirements for the various dwelling units to be found in one location instead of needing to refer to Chapter 6 and then finding out that an exception allows either the elimination or the revision of the grab bar requirements.

The relocation of the swing-up grab bar provisions from Section 604.5.3 provides an example of how this should also help clarify the application of these requirements. The swing-up grab bar requirements were and are only applicable to Type B dwelling units. However, because of the language of the standard and because these requirements were found within the general grab bar provisions of Section 604.5, it often led to misapplication or interpretation of the requirements and swing-up grab bars being installed or specified in public toilet rooms or Accessible and Type A dwelling units.

See the pages covering Sections 1003.11 and 1004.11, and the general page for Chapter 10, to see the new location of these requirements and for additional related discussion.

40

604.7, 604.11.7 continued

requirements, the A117 committee did not believe that having two measuring points (the back wall and the front of the fixture) was necessary and felt it would end up being overly restrictive. It also defeats the intent of eliminating measuring from the front of the water closet, which was a variable dimension based on the selection of the fixture. The A117 committee's belief was that the provisions of the A117.1 standard provided a more detailed and substantiated location for the dispenser and that it could at the least be considered as an alternate method of compliance and as providing equivalent accessibility to that specified by the ADA and ABA AG.

CHANGE TYPE: Modification

CHANGE SUMMARY: The location of toilet compartment doors is now regulated in a new table. The change will allow additional design options where the compartment is larger than the minimum required size.

604.9.3
Doors for Wheelchair-Accessible Compartments

2009 STANDARD:

604.9.3 Doors. Toilet compartment doors, including door hardware, shall comply with Section 404.1, except if the approach is to the latch side of the compartment door clearance between the door side of the stall and any obstruction shall be 42 inches (1065 mm) minimum. ~~Doors shall be located in the front partition or in the side wall or partition farthest from the water closet. Where located in the front partition, the door opening shall be 4 inches (100 mm) maximum from the side wall or partition farthest from the water closet. Where located in the side wall or partition, the door opening shall be 4 inches (100 mm) maximum from the front partition.~~ The door shall be self-closing. A door pull complying with Section 404.2.6 shall be placed on both sides of the door near the latch. Toilet compartment doors shall not swing into the required minimum area of the compartment.

604.9.3.1 Door Opening Location. The farthest edge of toilet compartment door opening shall be located in the front wall or partition or in the side wall or partition as required by Table 604.9.3.1.

CHANGE SIGNIFICANCE: Where water closet compartments are built to the minimum size required, this new table will not result in any change of the door location. However, where the compartments are built larger than the required minimum size, the new table and provisions will provide more design flexibility in establishing the location of the door. For larger compartments the standard will permit the door to be located based on either of two separate requirements. The standard previously required that the farthest edge of the door be located a maximum of 4 inches from

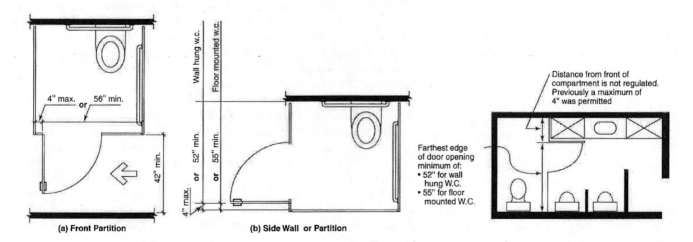

(a) Front Partition

(b) Side Wall or Partition

Farthest edge of door opening minimum of:
• 52" for wall hung W.C.
• 55" for floor mounted W.C.

Distance from front of compartment is not regulated. Previously a maximum of 4" was permitted

604.9.3 continues

604.9.3 continued   **TABLE 604.9.3.1   Door Opening Location**

Door Opening Location	Measured From	Dimension
Front Wall or Partition	From the side wall or partition closest to the water closet	56 inches (1420 mm) minimum
	or	
	From the side wall or partition farthest from the water closet	4 inches (100 mm) maximum
Side Wall or Partition Wall-Hung Water Closet	From the rear wall	52 inches (1320 mm) minimum
	or	
	From the front wall or partition	4 inches (100 mm) maximum
Side Wall or Partition Floor-Mounted Water Closet	From the rear wall	55 inches (1395 mm) minimum
	or	
	From the front wall or partition	4 inches (100 mm) maximum

the corner of the compartment farthest from the water closet, regardless of how large the compartment was.

The additional permitted dimensions for the opening location provide the same distance from the rear wall or side wall or partition as was previously required for either a floor-mounted or wall-hung water closet. For an opening located in the front partition, using the required minimum width of the compartment of 60 inches, the farthest edge of the door opening would be either 4 inches from the side wall or partition farthest from the water closet or 56 inches from the side wall or partition that is closest to the water closet. If the width of the compartment was increased to 72 inches, the new table would provide a range of 12 inches in which the farthest edge of the door could be located. This additional range of location may assist in being able to design around some other fixture, allow a larger support member for the partition, or provide some other benefit.

The 56-, 52-, and 55-inch dimensions that are shown in the table were derived from subtracting the 4-inch requirement from the minimum size for a water closet compartment. For example, a floor-mounted water closet would require a 59-inch minimum compartment depth based on the requirements of Section 604.9.2. Therefore the standard will permit the door to be located in the side partition either a maximum of 4 inches from the front partition or 55 inches (59 − 4 = 55) from the rear wall. Those two points will remain set regardless of the actual depth of the compartment.

Permitting the farthest edge of the opening to be located at a point equal to or greater than that required for a minimum size compartment allows greater flexibility in design and use of building space without lessening the current minimum accessibility criteria for entering and exiting a wheelchair-accessible compartment. In addition, a larger compartment will provide greater maneuvering space within the compartment and therefore make the compartment and access to the door easier for the users.

CHANGE TYPE: Modification

CHANGE SUMMARY: Change specifies that the clear floor space for a parallel approach to the sink or lavatory must be "centered" on the fixture.

606.2

Clear Floor Space for Lavatories and Sinks

2009 STANDARD:

606.2 Clear Floor Space. A clear floor space complying with Section 305.3, positioned for forward approach, shall be provided. Knee and toe clearance complying with Section 306 shall be provided. The dip of the overflow shall not be considered in determining knee and toe clearances.

Exceptions:

1. A parallel approach complying with Section 305, <u>and centered on the sink,</u> shall be permitted to a kitchen sink in a space where a cook top or conventional range is not provided.

2. The requirement for knee and toe clearance shall not apply to a lavatory in a toilet ~~and~~ <u>or</u> bathing facility for a single occupant, accessed only through a private office and not for common use or public use.

3. A knee clearance of 24 inches (610 mm) minimum above the floor shall be permitted at lavatories and sinks used primarily by children ages 6 through 12 where the rim or counter surface is 31 inches (785 mm) maximum above the floor.

4. A parallel approach complying with Section 305 <u>and centered on the sink,</u> shall be permitted at lavatories and sinks used primarily by children ages 5 and younger.

5. The requirement for knee and toe clearance shall not apply to more than one bowl of a multibowl sink.

6. A parallel approach <u>complying with Section 305 and centered on the sink,</u> shall be permitted at wet bars.

CHANGE SIGNIFICANCE: Adding the additional detail regarding the clear floor space location in relation to the sink helps provide additional

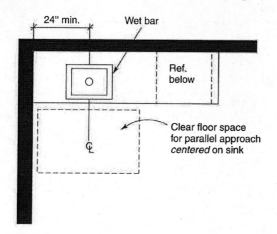

606.2 continues

606.2 continued

clarity, eliminate potential debate regarding the intent, and makes the sinks more likely to be usable. The new text dealing with the parallel approach options clarify that the sink needs to be located so that the clear floor space can be "centered on the sink" instead of simply being adjacent to any part of it.

This revision does have the potential to affect the location of the sink but it will also ensure that the clear floor space is located so that the sink is within reach from the floor space. To illustrate this change, compare the text of Exception 6 in the previous edition with that of the new edition. Under the 2003 edition the sink at the wet bar in a Type B dwelling unit could have been located all the way to the end of a cabinet so that the sink was placed in a corner. This new text would require that the bar sink be moved so that it was centered at least 24 inches from the adjacent wall so that the 48-inch clear floor space would be centered on the sink. In addition, under the previous language, if a designer could show that the clear floor space aligned with any portion of the sink it could have been argued that the design complied with the standard. Obviously, such a layout would not have met the intent of the provision, which is to make sure that the sink is accessible and usable from the clear floor space.

The revision in Exception 2 is similar to a change in Sections 603.2.2 and 606.3. See the change for Section 603.2 (see portion for Section 603.2.2) that was addressed earlier for additional discussion. This revision clarifies that the provision applies to either a toilet *or* a bathing facility and that you do not need to have both items in order to use the exception.

CHANGE TYPE: Modification

CHANGE SUMMARY: The height of the lower grab bar on the back wall has been modified to permit a range versus a specific height. The control end wall grab bar provisions are modified for consistency.

607.4.1.1, 607.4.2.1
Grab Bars at Bath Tubs

2009 STANDARD:

607.4.1 Bathtubs with Permanent Seats. For bathtubs with permanent seats, grab bars complying with Section 607.4.1 shall be provided.

607.4.1.1 Back Wall. Two horizontal grab bars shall be provided on the back wall, one complying with Section 609.4 and the other ~~9 inches (230 mm)~~ located 8 inches (205 mm) minimum and 10 inches (255 mm) maximum above the rim of the bathtub. Each grab bar shall be located 15 inches (380 mm) maximum from the head end wall and extend to 12 inches (305 mm) maximum from the control end wall.

607.4.1.2 Control End Wall. Control end wall grab bars shall comply with Section 607.4.1.2.

Exception: An L-shaped continuous grab bar of equivalent dimensions and positioning shall be permitted to serve the function of separate vertical and horizontal grab bars.

607.4.1.2.1 Horizontal Grab Bar. A horizontal grab bar 24 inches (610 mm) minimum in length shall be provided on the control end wall ~~at the front edge of the bathtub~~ beginning near the front edge of the bathtub and extending toward the inside corner of the bathtub.

607.4.1.2.2 Vertical Grab Bar. A vertical grab bar 18 inches (455 mm) minimum in length shall be provided on the control end wall 3 inches (75 mm) minimum ~~to~~ and 6 inches (150 mm) maximum above the horizontal grab bar, and 4 inches (100 mm) maximum inward from the front edge of the bathtub.

607.4.2 Bathtubs without Permanent Seats. For bathtubs without permanent seats, grab bars complying with Section 607.4.2 shall be provided.

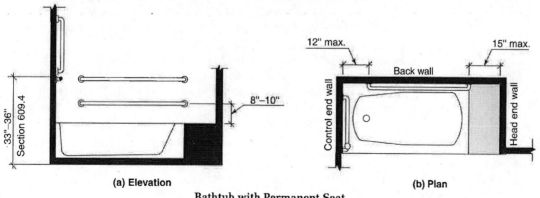

(a) Elevation (b) Plan

Bathtub with Permanent Seat

607.4.1.1, 607.4.2.1 continues

607.4.1.1, 607.4.2.1 continued

607.4.2.1 Back Wall. Two horizontal grab bars shall be provided on the back wall, one complying with Section 609.4 and the other ~~9 inches (230 mm)~~ located 8 inches (205 mm) minimum and 10 inches (255 mm) maximum above the rim of the bathtub. Each grab bar shall be 24 inches (610 mm) minimum in length, located 24 inches (610 mm) maximum from the head end wall and extend to 12 inches (305 mm) maximum from the control end wall.

607.4.2.2 Control End Wall. Control end wall grab bars shall comply with Section ~~607.4.2.2~~607.4.1.2.

~~**Exception:** An L-shaped continuous grab bar of equivalent dimensions and positioning shall be permitted to serve the function of separate vertical and horizontal grab bars.~~

~~**607.4.2.2.1 Horizontal Grab Bar.** A horizontal grab bar 24 inches (610 mm) minimum in length shall be provided on the control end wall beginning near the front edge of the bathtub and extend toward the inside corner of the bathtub.~~

~~**607.4.2.2.2 Vertical Grab Bar.** A vertical grab bar 18 inches (455 mm) minimum in length shall be provided on the control end wall 3 inches (76 mm) minimum to 6 inches (150 mm) maximum above the horizontal grab bar, and 4 inches (102 mm) maximum inward from the front edge of the bathtub.~~

CHANGE SIGNIFICANCE: The revision within Sections 607.4.1.1 and 607.4.2.1 provide a range for the installation of the lower grab bar on the back wall versus the absolute dimension that was found in the previous edition. This will help make compliance and enforcement easier by providing a range and eliminates questions where the bar was slightly above or below the previously specified 9-inch dimension. This revision also helps to coordinate the A117.1 standard with the requirements found within the ADA and ABA AG.

The changes within Sections 607.4.1.2.1 and 607.4.2.2.1 dealing with the grab bars on the control end of the tub are modified so the requirements are consistent regardless of the type of seat used on the tub. Besides making the requirements consistent, the change in Section 607.4.2.2.1 will ensure that the two sections remain coordinated and will eliminate the likelihood that they would become inconsistent again. The revised text that is used in Section 607.4.1.2.1 matches what was found within 607.4.2.2.1 of the 2003 standard. One slight change that results from this is that instead of requiring the grab bar to start "*at* the front edge of the bathtub," it is allowed to start "*near* the front edge of the bathtub." For people who try to be code literal in their design or enforcement, it is practically impossible to begin the bar immediately "at" the front edge when most bars have mounting brackets and cover plates that extend beyond the point where the bar begins.

CHANGE TYPE: Modification

CHANGE SUMMARY: Roll-in showers will now require seats. Permanent non-folding seats will be permitted in showers larger than the minimum dimensions. Revisions also include a format change to coordinate the three sections dealing with the various types of shower compartments. In addition, the seat requirements have been moved into each of the individual shower compartment sections.

608.2

Sizes and Clearances for Shower Compartments

2009 STANDARD:

608.2 Size, and Clearances and Seat. Shower compartments shall have sizes, clearances, and seats complying with Section 608.2.

608.2.1 Transfer-Type Shower Compartments.   Transfer-type shower compartments shall comply with Section 608.2.1.

608.2.1.1 Size.   Transfer-type shower compartments shall have a clear inside dimension of 36 inches (915 mm) in width and 36 inches (915 mm) in depth, measured at the center point of opposing sides. An entry 36 inches (915 mm) minimum in width shall be provided.

608.2.1.2 Clearance.   A clearance of 48 inches (1220 mm) minimum in length measured perpendicular from the control wall, and 36 inches (915 mm) minimum in depth shall be provided adjacent to the open face of the compartment.

608.2.1.3 Seat.   A folding or non-folding seat complying with Section 610 shall be provided on the wall opposite the control wall.

Exception: A seat is not required to be installed in a shower for a single occupant, accessed only through a private office and not for common use or public use, provided reinforcement has been installed in walls and located so as to permit the installation of a shower seat.

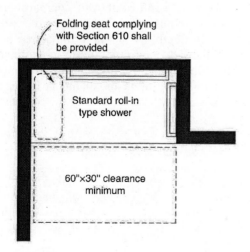

Folding seat complying with Section 610 shall be provided

Standard roll-in type shower

60"×30" clearance minimum

608.2 continues

608.2 continued

608.2.2 Standard Roll-in-Type Shower Compartments. <u>Standard roll-in-type shower compartments shall comply with Section 608.2.2.</u>

608.2.2.1 Size. Standard roll-in-type shower compartments shall have a clear inside dimension of 60 inches (1525 mm) minimum in width and 30 inches (760 mm) minimum in depth, measured at the center point of opposing sides. An entry 60 inches (1525 mm) minimum in width shall be provided.

608.2.2.2 Clearance. A clearance of 60 inches (1525 mm) minimum in length adjacent to the 60-inch (1525 mm) width of the open face of the shower compartment, and 30 inches (760 mm) minimum in depth, shall be provided.

Exception: A lavatory complying with Section 606 shall be permitted at the end of the clearance opposite the <u>seat.</u> ~~shower compartment side where shower controls are positioned. Where shower controls are located on the back wall and no seat is provided, the lavatory shall be permitted at either end of the clearance.~~

608.2.2.3 Seat. <u>A folding seat complying with Section 610 shall be provided on an end wall.</u>

Exceptions:

1. <u>A seat is not required to be installed in a shower for a single occupant accessed only through a private office and not for common use or public use, provided reinforcement has been installed in walls and located so as to permit the installation of a shower seat.</u>
2. <u>A fixed seat shall be permitted where the seat does not overlap the minimum clear inside dimension required by Section 608.2.2.1.</u>

608.2.3 Alternate Roll-in-Type Shower Compartments. <u>Alternate roll-in-type shower compartments shall comply with Section 608.2.3.</u>

608.2.3.1 Size. Alternate roll-in shower compartments shall have a clear inside dimension of 60 inches (1525 mm) minimum in width, and 36 inches (915 mm) in depth, measured at the center point of opposing sides. An entry 36 inches (915 mm) minimum in width shall be provided at one end of the 60-inch (1525 mm) width of the compartment. A seat wall, 24 inches (610 mm) minimum and 36 inches (915 mm) maximum in length, shall be provided on the entry side of the compartment.

608.2.3.2 Seat. <u>A folding seat complying with Section 610 shall be provided on the seat wall opposite the back wall.</u>

Exception: <u>A seat is not required to be installed in a shower for a single occupant, accessed only through a private office and not for common use or public use, provided reinforcement has been installed in walls and located so as to permit the installation of a shower seat.</u>

~~**608.4 Seats.** A folding or nonfolding seat shall be provided in transfer-type shower compartments. A seat shall be provided in an alternate roll-in-type shower compartment. In standard and alternate roll-in-type showers where a seat is provided, if the seat extends over the minimum clear inside dimension required by Section 608.2.2 or 608.2.3, the seat shall be a folding seat. Seats shall comply with Section 610.~~

Exceptions:

1. ~~A shower seat is not required to be installed in a shower facility for a single occupant, accessed only through a private office and not for common use or public use, provided reinforcement has been installed in walls and located so as to permit the installation of a shower seat complying with Section 608.4.~~

2. ~~In Type A units, a shower seat is not required to be installed where reinforcement complying with Section 1003.11.4 is installed for the future installation of a shower seat.~~

610.3 Shower Compartment Seats. ~~Where a seat is provided in a standard roll-in shower compartment, it shall be a folding type and shall be on the wall adjacent to the controls.~~ The height of ~~the~~ <u>shower compartment</u> seat<u>s</u> shall be 17 inches (430 mm) minimum and 19 inches (485 mm) maximum above the bathroom floor, measured to the top of the seat. In transfer-type and alternate roll-in-type showers, the seat shall extend along the seat wall to a point within 3 inches (75 mm) of the compartment entry. In standard roll-in-type showers, the seat shall extend from the control wall to a point within 3 inches (75 mm) of the compartment entry. Seats shall comply with Section 610.3.1 or 610.3.2.

CHANGE SIGNIFICANCE: The most apparent revision of these provisions is the format change to coordinate the three sections dealing with the various types of shower compartments. In addition, the seat requirements have been moved from where they had previously been in Section 608.4 into each of the individual shower compartment sections. This will make the seat requirements easier to determine and ensure they are not overlooked.

Perhaps the biggest revision is found in the seat requirements for roll-in showers, where the general provisions will typically require the installation of a folding seat instead of simply regulating the seat if it is provided. Many wheelchair users transfer to a shower seat when bathing because they do not have a separate wheelchair that can be used in the shower. The addition of the seat for the roll-in shower will make all of the types of bathing facilities more consistent by requiring a seat and should make the roll-in showers usable for a wider range of individuals with different mobility aids. Providing a folding or fixed seat within the shower compartment will ensure that a seat is always available for those who need it. One of the problems with a removable seat or separate shower chair is that they often are not available or not available in adequate numbers when and where they are needed.

The requirement for a folding seat in a roll-in shower ensures that the shower can be used as a roll-in shower by not having the floor space in the shower obstructed. Exception 2 in Section 608.2.2.3 will allow a fixed

608.2 continues

608.2 continued   seat in the roll-in shower provided the shower size is increased so the permanent seat does not overlap the space needed for a roll-in.

The revision in the general seating requirements of Section 610.3 coordinates with the change made in Section 608.2.2.3, where the seat is now required instead of an option that is regulated only when it is provided. See also the discussions for Sections 608.3.2 and 608.4.2 that follow later in this book.

CHANGE TYPE: Modification

CHANGE SUMMARY: Grab bar requirements have been modified because of the fact that a seat is required by the new standard. Additional detail is provided for the grab bar requirements and locations.

608.3.2
Grab Bars for Standard Roll-in-Type Showers

2009 STANDARD:

608.3.2 Standard Roll-in-Type Showers. In standard roll-in-type showers, ~~grab bars shall be provided on three walls of showers without seats. Where a seat is provided in a standard roll-in-type shower,~~ a grab bars shall be provided on the back wall ~~and on the wall opposite~~ beginning at the edge of the seat. The grab bars shall not be provided above the seat. The back wall grab bar shall extend the length of the wall but shall not be required to exceed 48 inches (1220 mm) in length. Where a side wall is provided opposite the seat within 72 inches (1830 mm) of the seat wall, a grab bar shall be provided on the side wall opposite the seat. The side wall grab bar shall extend the length of the wall but shall not be required to exceed 30 inches (760 mm) in length. Grab bars shall be 6 inches (150 mm) maximum from the adjacent wall.

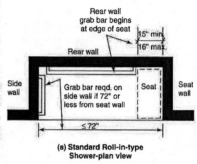

(a) Standard Roll-in-type Shower-plan view

CHANGE SIGNIFICANCE: Because roll-in-type showers are now required to be provided with a folding seat, the previous language dealing with showers without seats is inappropriate. Therefore the language of this section was modified to remove the text related to showers without seats and to provide additional detail for the grab bar requirements.

Primarily the added text will address situations where the shower size exceeds the minimum required size from Section 608.2.2.1. The most notable of these changes is that the length of the grab bar on the back wall is not required to exceed 48 inches in length and a grab bar is not required on the end wall if that end wall is more than 72 inches from the seat wall. This 48-inch maximum length would be applicable when the length of the shower exceeds the minimum required dimension of 60 inches due to the fact that the grab bar is intended to start at the front edge of the seat and is not to be provided above the seat. Therefore, depending on the type of seat (rectangular or "L" shaped) and the requirements of Section 610.3, a 48-inch grab bar would only be permitted if the shower width was large enough (63 or 70 inches minimum) to permit the bar and yet not extend over the seat. If the bar does not begin immediately at the end wall and begins as far as the permitted 6 inches away, then the shower compartment width would need to be even larger (69 or 76 inches minimum) before the 48-inch grab bar limit would be applicable.

Where the shower compartment size exceeds 72 inches the standard will eliminate the requirement for the end wall grab bar. Where the shower compartment reaches such a large size, the grab bar would be beyond the reach of a person using the seat or seated in a wheelchair near the opposite end. This will help designers trying to provide an accessible shower in a gang-shower room.

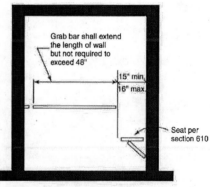

(b) Elevation of back wall

608.4.1

Controls and Hand Showers for Transfer-Type Showers

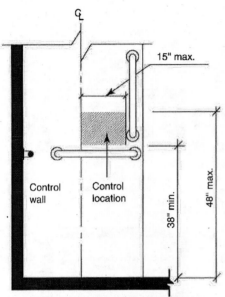

Control wall

Control location

15" max.

38" min.

48" max.

CHANGE TYPE: Modification

CHANGE SUMMARY: This section was reformatted and item 3 is modified so that the controls are located toward the opening of the shower.

2009 STANDARD:

608.4 Controls and Hand Showers. Controls and hand showers shall comply with Sections 608.4 and 309.4.

608.4.1 Transfer-Type Showers. In transfer-type showers, the controls and hand shower shall be located:

1. On the control wall opposite the seat;
2. At a height of 38 inches (965 mm) minimum and 48 inches (1220 mm) maximum above the shower floor, within and
3. 15 inches (380 mm), left or right, of maximum, from the centerline of the seat control wall toward the shower opening.

CHANGE SIGNIFICANCE: The revision of the text in the new item 3 will place the controls closer to the opening and ensure that they are reachable from outside of the compartment. The standard previously allowed the controls to be located either to the "left or right of the centerline of the seat." This ensured that the controls were located within the reach of someone using the seat but if the controls were located toward the back wall they may not have been within the reach of a person who was outside of the shower. By revising the text to require the controls be either at the centerline or toward the shower opening, it will keep the controls within reach of the seat but will allow a person to operate and adjust them from the outside of the compartment prior to entering the shower. This essentially reduces the space where the controls can be located by one-half but it will make control from outside of the compartment easier.

This revision will also help to coordinate the requirements between the A117.1 standard and those of the ADA and ABA AG. This will provide additional consistency and eliminate confusion regarding which document should be followed.

CHANGE TYPE: Modification

CHANGE SUMMARY: Changes to require a seat in a roll-in shower resulted in coordinating revisions to the location of shower controls and greatly reduced the options that existed previously. The controls must be located in a much smaller area than previously permitted.

608.4.2
Controls and Hand Showers for Standard Roll-in Showers

2009 STANDARD:

608.4.2 Standard Roll-in Showers. In standard roll-in showers, the controls and hand shower shall be located <u>on the back wall</u> ~~38 inches (965 mm) minimum and~~ <u>above the grab bar,</u> 48 inches (1220 mm) maximum above the shower floor. ~~In standard roll-in showers with seats, the controls and hand shower shall be located on the back wall,~~ <u>and</u> ~~no more than~~ <u>16 inches (405 mm) minimum and</u> 27 inches (685 mm) maximum from the end wall behind the seat.

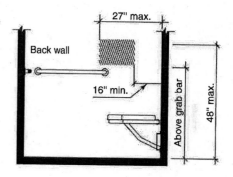

CHANGE SIGNIFICANCE: Some of the changes in this section are necessary due to the revisions in Section 608.2.2.3, which require the installation of a folding seat in a roll-in shower. Previously the standard did not require the installation of the seat and therefore this section provided options as to where the controls could be located when a seat was not provided. (For additional related discussion, see the page covering Section 608.2 earlier in this book.)

Under the previous standard, where a seat was not installed, the controls could go on any wall of the shower. Where a seat was installed the controls were limited to the back wall of the shower. Because a seat is now required, the control location is limited to the back wall as previously done but the location has been moved so that it is a minimum of 16 inches from the end wall behind the seat. Requiring a minimum 16-inch dimension will essentially move the controls so that they are located near the front of or in front of the seat and not above the seat. Moving the controls forward from the seat wall should help to assure access to the controls both from the seat and when the shower is used as a roll-in. If the controls were not moved forward and a fixed seat was installed (as permitted by 608.2.2.3 Exception 2), it would have the potential to create an obstruction to the controls.

The 38-inch minimum height for the controls was eliminated both to coordinate with the ADA and ABA AG and to ensure the controls do not conflict with the grab bar. Grab bars are permitted to be installed 33 inches minimum and 36 inches maximum in height. Eliminating this 38-inch minimum requirement will allow the controls to be located based on the requirements of Section 609.3 Exception 1 and located at a lower height where the grab bar is installed at the lower limits of the permitted range.

608.5

Hand Showers

CHANGE TYPE: Modification

CHANGE SUMMARY: The requirements for hand held showers and the option of using a fixed shower head at a 48-inch height have been revised and clarified.

2009 STANDARD:

608.4.3 Alternate Roll-in Showers. In alternate roll-in showers, the controls and hand shower shall be located 38 inches (965 mm) minimum and 48 inches (1220 mm) maximum above the shower floor. In alternate roll-in showers with controls and hand shower located on the end wall adjacent to the seat, the controls and hand shower shall be 27 inches (685 mm) maximum from the seat wall. In alternate roll-in showers with the controls and hand shower located on the back wall opposite the seat, the controls and hand shower shall be located within 15 inches (380 mm), left or right, of the centerline of the seat.

Exception: A fixed shower head with the controls and shower head located on the back wall opposite the seat shall be permitted.

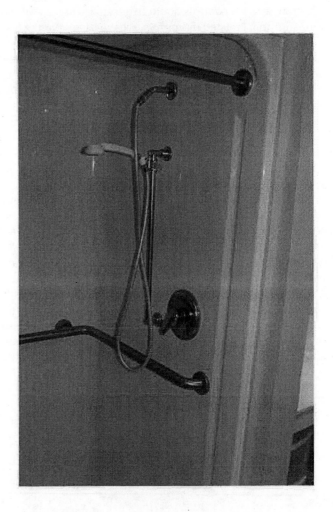

608.5 Hand Showers. A hand shower with a hose 59 inches (1500 mm) minimum in length, that can be used both as a fixed shower head and as a hand shower, shall be provided. The hand shower shall have a control with a nonpositive shut-off feature. <u>Where provided, an</u> adjustable-height <u>hand</u> shower ~~head~~ mounted on a vertical bar shall be installed so as to not obstruct the use of grab bars.

Exception: <u>In other than Accessible units and Type A units, a</u> fixed shower head <u>located 48 inches (1220 mm) maximum above the shower floor</u> shall be permitted in lieu of a hand shower ~~where the scoping provisions of the administrative authority require a fixed shower head.~~

CHANGE SIGNIFICANCE: The deletion of the exception in Section 608.4.3 simply removes language that is not needed. The exception within Section 608.5 permits a fixed shower head and can be used instead of the deleted exception in 608.4.3. One difference, whether intended or not, is that under the 2003 edition of the standard the fixed shower head and controls in an alternate roll-in shower had to be located on the back wall. When using the new standard and the exception in Section 608.5, there is nothing that would prohibit the fixed shower head from being located on the end wall adjacent to the seat as is permitted in Section 608.4.3 for the hand shower and controls.

By adding the phrase "where provided" in Section 608.5, it clarifies that an adjustable-height hand shower is not required. As previously written, this sentence with the phrasing "shall be installed" was occasionally misinterpreted as requiring the hand shower to be adjustable. The "shall be installed" phrase needed to be read in the context that it "shall be installed *so as to not obstruct the use of grab bars.*" A similar change can also be found in Section 607.6 and applies to bathtubs.

The exception in Section 608.5 is modified to provide several new limitations. First a fixed shower head that replaces a hand shower is limited to a maximum height of 48 inches. This height corresponds to the height requirement for the hand shower in the various subsections of 608.4. This also ensures the shower head is within reach so that it can be adjusted by the user. This height limit is therefore not only practical but also coordinates with the requirements of the ADA and ABA AG. Second, the provision is modified so that the option for a fixed shower is permitted "in other than Accessible units and Type A units" and is not limited to "where the scoping provisions of the administrative authority require a fixed shower head." These revisions were primarily done to coordinate with the ADA and ABA AG. The federal law permits a fixed shower head in lieu of a hand shower in facilities that are not medical care facilities, long-term care facilities, transient lodging guest rooms, or residential dwelling units. Therefore the exclusion for Accessible and Type A units will correspond to many of those uses. While this exception is no longer tied to "where the scoping provisions of the administrative authority require a fixed shower head," if the jurisdiction does want to require a fixed shower in lieu of a hand shower, the jurisdiction can continue to do that through its adopting process and Section 201.

609

Grab Bar Clearance and Height

CHANGE TYPE: Modification

CHANGE SUMMARY: The grab bar height requirements are reformatted to make the children's requirements more apparent. Provisions are added to address recessed toilet paper dispensers and their location in relationship to the clear spaces behind the grab bars.

2009 STANDARD:

609.3 Spacing. The space between the wall and the grab bar shall be 1-1/2 inches (38 mm). The space between the grab bar and projecting objects below and at the ends of the grab bar shall be 1-1/2 inches (38 mm) minimum. The space between the grab bar and projecting objects above the grab bar shall be 12 inches (305 mm) minimum.

Exceptions:

1. The space between the grab bars and shower controls, shower fittings, and other grab bars above the grab bar shall be permitted to be 1-1/2 inches (38 mm) minimum.

2. ~~Swing-up grab bars shall not be required to comply with Section 609.3.~~

2. <u>Recessed dispensers projecting from the wall 1/4 inch (6.3 mm) maximum measured from the face of the dispenser and complying with Section 604.7 shall be permitted within the 12-inch (305-mm) space above and the 1-1/2 inch (38-mm) spaces below and at the ends of the grab bar.</u>

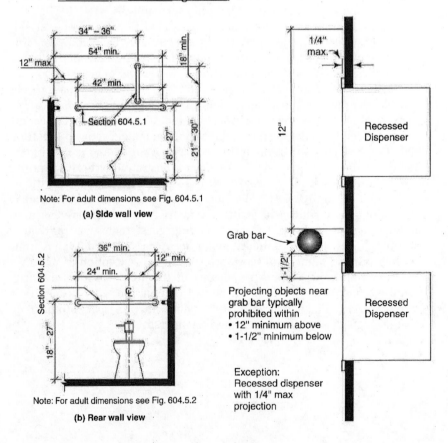

Note: For adult dimensions see Fig. 604.5.1

(a) Side wall view

Note: For adult dimensions see Fig. 604.5.2

(b) Rear wall view

Projecting objects near grab bar typically prohibited within
• 12" minimum above
• 1-1/2" minimum below

Exception: Recessed dispenser with 1/4" max projection

609.4 Position of Grab Bars.

609.4.1 General. Grab bars shall be installed in a horizontal position, 33 inches (840 mm) minimum and 36 inches (915 mm) maximum above the floor measured to the top of the gripping surface <u>or shall be installed as required by Items 1 through 3.</u> At water closets primarily for children's use complying with Section 604.10, grab bars shall be installed in a horizontal position 18 inches (455 mm) minimum to 27 inches (685 mm) maximum above the floor measured to the top of the gripping surface.

Exceptions:

1. The lower grab bar on the back wall of a bathtub required by <u>shall comply with</u> Sections 607.4.1.1 or 607.4.2.1.
2. Vertical grab bars required by <u>shall comply with</u> Sections 604.5.1, 607.4.1.2.2, 607.4.2.2.2, and 608.3.1.2.
3. <u>Grab bars at water closets primarily for children's use shall comply with Section 609.4.2.</u>

609.4.2 Position of Children's Grab Bars. <u>At water closets primarily for children's use complying with Section 604.11, grab bars shall be installed in a horizontal position 18 inches (455 mm) minimum and 27 inches (685 mm) maximum above the floor measured to the top of the gripping surface. A vertical grab bar shall be mounted with the bottom of the bar located between 21 inches (533 mm) minimum and 30 inches (760 mm) maximum above the floor and with the centerline of the bar located between 34 inches (865 mm) minimum and 36 inches (915 mm) maximum from the rear wall.</u>

609.7 Installation and Configuration. Grab bars shall be installed in any manner that provides a gripping surface at the locations specified in this standard and does not obstruct the clear floor space. <u>Horizontal and vertical grab bars shall be permitted to be separate bars, a single piece bar, or combination thereof.</u>

CHANGE SIGNIFICANCE: Numerous changes were made to the grab bar clearance, height, and configuration provisions but predominately they do not result in technical changes. The two technical changes in these sections can be found in the new Exception 2 in Section 609.3 dealing with recessed dispensers and in the fact that the children's grab bar provisions of Section 609.4.2 provide specific requirements for a vertical side wall grab bar.

The exceptions to the spacing requirements of Section 609.3 have been modified by the deletion of the previous Exception 2 and the addition of a new exception that addresses recessed dispensers. The exception dealing with swing-up grab bars was removed from this section and is covered in the Type B dwelling unit requirements of Section 1004.11. Since swing-up grab bars are only permitted in the Type B units, it is appropriate to have all of their requirements located in that section. See the page earlier in this book dealing with Sections 604.5 and 607.4 and the page at the start of Chapter 10 for additional related information.

609 continues

609 continued

The new Exception 2 in Section 609.3 was developed as a part of the changes addressing the location of toilet paper dispensers in Section 604.7. As a part of those changes, the relationship of the dispenser and the grab bar was an important aspect, and the dispenser location was tied to the clearances of Section 609.3. This new exception addresses the use of recessed dispensers by allowing them to project a maximum of 1/4 inch from the wall when they are located within the 12-inch space above or the 1-1/2-inch space that is required below or at the end of grab bars. The use of recessed fixtures provides additional options for the possible location of the dispensers by permitting them into the previously required clear spaces above and below the bar. Specifically addressing the recessed dispensers provides better understanding of how they should be regulated versus not addressing them and dealing with the resulting uncertainty and inconsistent enforcement. While not specifically addressed, it is the author's opinion that the intent was to also allow recessed waste disposal containers and/or seat cover dispensers. Manufacturers do make some products that are combined with the toilet paper dispensers or designed to work as a group of elements provided for the user.

The changes within Section 609.4 will primarily relocate the grab bar provisions that are applicable to water closets primarily for children's use into a new subsection (Section 609.4.2). The one technical difference is that the height and the location for the vertical grab bar have been added. The 2003 edition of the standard did not clearly indicate whether the vertical bar was required in this situation, and if it was, exactly where the bar was to be located.

The layout of Section 609.4 was also modified so that the "exceptions" are now references to other requirements. Due to the nature of these items and the code text, these items were truly not exceptions to the base paragraph's requirements. In addition, if they were exceptions then the users could elect to ignore them and not have to comply with the sections that they intend to require compliance with.

The revision in Section 609.7 inserts text that was previously found in Section 608.3.1.3 of the 2003 edition of the standard. Conceptually this goes with the previous text of Section 609.7 that allowed the bars to be installed "in any manner" that provided a gripping surface at the required locations. By placing this added text into this general grab bar section, it will allow for a continuous grab bar in showers, tubs, and around water closets. When the provision was previously located within Section 608.3.1.3 it was limited in application to transfer-type showers.

CHANGE TYPE: Modification

CHANGE SUMMARY: The clear floor space access to washers and dryers has been modified depending upon whether the machine is a front- or top-loading machine. The minimum height for front-loading machines has been increased for all but Type B dwelling units.

611

Height of and Approach to Washing Machines and Clothes Dryers

2009 STANDARD:

611 Washing Machines and Clothes Dryers

611.1 General. Accessible washing machines and clothes dryers shall comply with Section 611.

611.2 Clear Floor Space. A clear floor space complying with Section 305, positioned for parallel approach, shall be provided. For top-loading machines, the clear floor space shall be centered on the appliance. For front-loading machines, the centerline of the clear floor space shall be offset 24 inches (610 mm) maximum from the centerline of the door opening.

611.4 Height. Top-loading machines shall have the door to the laundry compartment 36 inches (915 mm) maximum above the floor. Front-loading machines shall have the bottom of the opening to the laundry compartment 15 inches (380 mm) minimum and 34 36 inches (865 915 mm) maximum above the floor.

1004 Type B Units

1004.10 Laundry Equipment. Washing machines and clothes dryers shall comply with Section 1004.10.

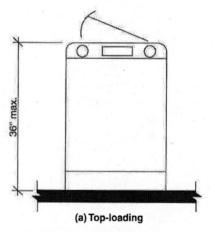

(a) Top-loading

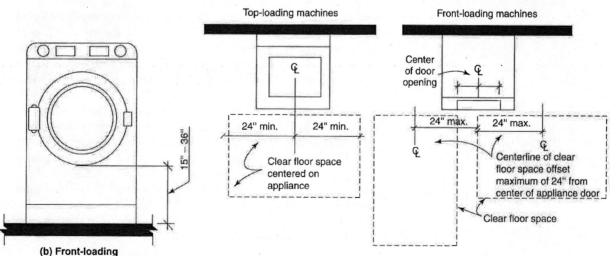

(b) Front-loading

611 continues

611 continued

1004.10.1 Clear Floor Space. A clear floor space complying with Section 305.3, ~~positioned for parallel approach,~~ shall be provided. ~~The clear floor space shall be centered on the appliance.~~ <u>A parallel approach shall be provided for a top-loading machine. A forward or parallel approach shall be provided for a front-loading machine.</u>

CHANGE SIGNIFICANCE: The clear floor space requirements have been modified to address top-loading and front-loading machines separately. Previously the standard did not make any distinction between the two types of machines. With more and more front-loading machines being sold on the market, it is important to recognize the differences between front- and top-loading machines. Front-loading washers and dryers, especially those with drawers underneath, provide a higher level of access. By revising this section the standard is providing more flexibility and addresses a type of machine that has not been previously addressed.

The revision in Section 611.2 will require the clear floor space to be centered on the machine as it previously was required in the 2003 edition of the standard, but now this requirement will apply only to top-loading machines. Front-loading washers and dryers will permit the clear floor space to be offset up to a maximum of 24 inches from the center of the door opening. Allowing the clear floor space for front-loading machines to be offset is similar to what the standard permits in a kitchen for the refrigerator, and it recognizes that positioning the space directly in front of the door may prevent the door from opening and may be detrimental to access.

The maximum height allowed for the opening into a front-loading laundry machine has been raised from 34 inches to 36 inches. The primary reason for this revision was to coordinate with Section 611.4 of the ADA and ABA AG. This revision would also appear to make sense from a practical standpoint. If the height of a top-loading machine that would require the user to reach the opening and then reach down into the equipment is permitted at 36 inches, then it would seem reasonable to also allow the front-loading machine with its easier access into the equipment to also be at that height.

The requirements for laundry equipment within a Type B dwelling unit have also been revised but will differ from the requirements of Section 611. Clearances at laundry equipment are not required by the Fair Housing Act (FHA). This means that the A117.1 standard exceeds the requirements of the FHA simply by requiring the clear floor space. In the Type B units a parallel approach is required for top-loading machines while a forward or parallel approach is acceptable for front-loading laundry equipment. In addition, the standard no longer requires that the clear floor spaces be centered on the equipment. These changes will provide more options to approach and use the appliances.

It is important that users recognize the differences between the Type B unit requirements and those of Section 611. By reference, the Accessible and Type A dwelling units would also follow the requirements of Section 611. The option for a forward approach to top-loading washers and dryers was not added into the standard in order that the requirements in Section 611 and those of the Accessible and Type A units could match the requirements of ADA and ABA AG Section 611.2. Therefore for Accessible and Type A units, the A117.1 standard is modified to match the federal ADA and ABA AG and will exceed the FHA.

CHANGE TYPE: Addition

CHANGE SUMMARY: A new section has been added to provide the technical requirements to make saunas and steam rooms accessible.

2009 STANDARD:

612
Saunas and Steam Rooms

612 Saunas and Steam Rooms

612.1 General. Saunas and steam rooms shall comply with Section 612.

612.2 Bench. Where seating is provided in saunas and steam rooms, at least one bench shall comply with Section 903. Doors shall not swing into the clear floor space required by Section 903.2.

612.3 Turning space. A turning space complying with Section 304 shall be provided within saunas and steam rooms.

CHANGE SIGNIFICANCE: The standard did not previously provide technical requirements for saunas or steam rooms. The addition of this section will provide these technical requirements within the A117.1 standard and will coordinate with the ADA and ABA AG.

The requirements are not especially onerous but will simply provide references to the bench requirements of Section 903, limit the potential for a door to swing into the required clear floor space serving the bench, and also require that a turning space complying with Section 304 is provided within the room. The bench requirements of Section 903 regulate the need for a clear floor space, address the size and height of the bench, require back support, regulate the structural strength, and state the need for slip-resistant surfaces on the seat. For information regarding changes affecting the bench, see the page dealing with Section 903 in this book.

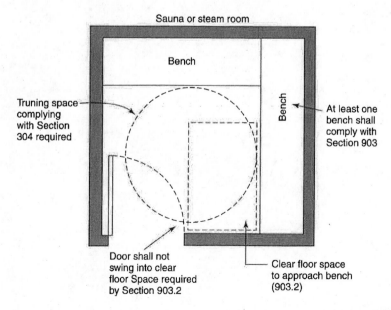

Sauna or steam room

Bench

Bench

Truning space complying with Section 304 required

At least one bench shall comply with Section 903

Door shall not swing into clear floor Space required by Section 903.2

Clear floor space to approach bench (903.2)

612 continues

612 continued   An item that was not addressed, which may be considered a safety issue or customer service issue, would be if the person using the sauna or a steam room would need to be able to move his or her chair out of the room him- or herself once he or she had transferred to the bench, or if it would be acceptable to have someone else remove and then bring the wheelchair back in. It would not be advisable to have a metal wheelchair remain in the hot or wet environment for an extended period of time.

CHANGE TYPE: Addition

CHANGE SUMMARY: Scoping requirements are added into the sign provisions to help clarify the type and requirements for signs to be considered as accessible. Depending upon the location and purpose of the sign, the standard specifies what requirements the sign must meet.

703.1

General Provisions for Signs

2009 STANDARD:

703 Signs

703.1 General.   Accessible signs shall comply with Section 703. <u>Tactile signs shall contain both raised characters and braille. Where signs with both visual and raised characters are required, either one sign with both visual and raised characters, or two separate signs, one with visual, and one with raised characters, shall be provided.</u>

<u>**703.1.1 Designations.**   Interior and exterior signs identifying permanent rooms and spaces shall comply with Sections 703.1, 703.2, and 703.3.</u>

<u>**Exception:**   Exterior signs that are not located at the door to the space they serve shall not be required to comply with Section 703.3.</u>

<u>**703.1.2 Directional and Informational Signs.**   Signs that provide direction to or information about interior spaces and facilities of the site shall comply with Section 703.2.</u>

<u>**703.1.3 Pictograms.**   Where pictograms are provided as designations of permanent interior rooms and spaces, the pictograms shall comply with Section 703.5 and shall have text descriptors located directly below the pictogram field and complying with Sections 703.2 and 703.3.</u>

703.1 continues

703.1 continued

Exception: Pictograms that provide information about a room or space, such as "no smoking," occupant logos, and the International Symbol of Accessibility, are not required to have text descriptors.

703.5.4 Text Descriptors. ~~Where text descriptors for pictograms are required, they shall be located directly below the pictogram field. Text descriptors shall comply with Sections 703.3 and 703.4.~~

CHANGE SIGNIFICANCE: The revisions to Chapter 7 and the new text in this section will help to clarify the types of signs that are required. Depending upon the purpose of the sign, the standard will require either visual or raised characters and in certain situations braille. These revisions will help clarify the signage requirements and also coordinate the A117.1 standard with the ADA and ABA AG signage provisions.

One very important detail that is found in the text of Section 703.1 is the requirement that "Tactile signs shall contain both raised characters and braille." Therefore when the sign is required to be "tactile" the requirements of both Sections 703.3 and 703.4 would be applicable. In addition, there are numerous places within the standard where the term "tactile" has been replaced with either the word "raised" or with the phrase "containing raised characters and braille." This distinction will be addressed later but it is important to be aware that there are numerous locations within the standard where the term "tactile" has been revised. This simple change in wording will help with the application and enforcement of the standard by clearly specifying what the specific provisions are for each type of sign. The requirements for visual characters are found in Section 703.2, the raised character requirements are in Section 703.3, and the braille provisions are in Section 703.4. So, depending on whether the sign has visual or raised characters required and whether or not braille is required, it should be much easier to determine what provisions apply to the sign.

The last sentence of Section 703.1 is also an important general provision to recognize since it will allow a single sign to include both visual and raised characters or would permit the option of installing two separate signs. The option of using two separate signs will allow one sign to meet the visual character requirements and the other sign to comply with the raised character provisions. For example, a room might have a visual sign on or over the door to make it easier for people to see at a distance or over crowds and then provide a separate raised and braille sign at the jamb for persons with vision impairments.

The provisions of section 703.1.3 help clarify that pictograms are regulated only for room designations and are not for items such as the International Symbol of Accessibility or items such as TTY devices. Although this is really a scoping issue that is typically handled by a separate document (see Section 201), the inclusion of this text helps address items that the building code official probably does not regulate and will help make the application of the standard's signage requirements easier to understand. This section and exception reflect the committee's belief that text descriptors are not needed for many pictograms. Because the requirements for pictogram text descriptors have been included in Section 703.1.3, the requirements that were previously found in Section 703.5.4 are no longer needed. Note that 703.1.3 is limited to room designations, not all pictograms. For example, the restroom designation in Figure 703.5

of the standard does have to meet the pictogram requirements as well as provide a text descriptor and braille. A pictogram used for other purposes, such as the international symbol for a volume-controlled telephone (see Figure 703.6.3.4) or the International Symbol for Accessibility (see Figure 703.6.3.1) used to designate an accessible entrance or an accessible parking space, does not have to meet pictogram requirements. Note that the exception would allow for the company logo to be on the front door of the office suite and not include a text descriptor.

Pictograms used for purposes other than room designations do not have a specified size requirement. However, good guidance might be the height for visual characters when provided at the same viewing distance. For example, the wheelchair symbol at the accessible entrance should probably be at least 3 inches tall if viewed from about 15 feet (see Table 703.2.4).

703.2.1

Visual Character Options for Signs

CHANGE TYPE: Modification

CHANGE SUMMARY: The revisions provide scoping guidance related to the various types of visual characters. This section also serves as the introduction to a new section of visual characters that are regulated under the variable message sign (VMS) signage requirements of Section 703.7.

2009 STANDARD:

703.2 Visual Characters.

703.2.1 General. ~~Visual characters shall comply with Section 703.2.~~ Visual characters shall comply with the following:

Exception: ~~Visual characters complying with Section 703.3 shall not be required to comply with Section 703.2.~~

1. Visual characters that also serve as raised characters shall comply with Section 703.3, or
2. Visual characters on VMS signage shall comply with Section 703.7, or
3. Visual characters not covered in items 1 and 2 shall comply with Section 703.2.

Exception: The visual and raised requirements of item 1 shall be permitted to be provided by two separate signs that provide corresponding information provided one sign complies with Section 703.2 and the second sign complies with Section 703.3.

CHANGE SIGNIFICANCE: This general section provides the scoping and direction to the various sections of the standard that apply to visual signs. This also shows that there are essentially three separate types of visual characters addressed by the standard. In general, visual characters are regulated by the provisions of Section 703.2. Where VMS signs are used, the characters are required to comply with the new requirements for VMS that are found in Section 703.7. In those situations where a visual character is also expected to serve as a raised character, then item 1 will direct the user to the provisions of Section 703.3.

The exception, which is applicable to item 1, is essentially a repeat of the requirements found at the end of section 703.1. This exception will direct the user to both the visual and the raised character requirements when a decision to provide two separate signs has been made.

CHANGE TYPE: Modification

CHANGE SUMMARY: A new exception has been added to address the character height and line spacing required in assembly seating areas where the viewing distance to the characters is 100 feet or greater.

2009 STANDARD:

703.2.4 Character Height. The uppercase letter "I" shall be used to determine the allowable height of all characters of a font. The uppercase letter "I" of the font shall have a minimum height complying with Table 703.2.4. Viewing distance shall be measured as the horizontal distance between the character and an obstruction preventing further approach towards the sign.

Exception: In assembly seating where the maximum viewing distance is 100 feet (30.5 m) or greater, the height of the uppercase "I" of fonts shall be permitted to be 1 inch (25 mm) for every 30 feet (9145 mm) of viewing distance, provided the character height is 8 inches (205 mm) minimum. Viewing distance shall be measured as the horizontal distance between the character and where someone is expected to view the sign.

703.2.8 Line Spacing. Spacing between the baselines of separate lines of characters within a message shall be 135 percent minimum and 170 percent maximum of the character height.

Exception: In assembly seating where the maximum viewing distance is 100 feet (30.5 m) or greater, the spacing between the baselines of separate lines of characters within a message shall be permitted to be 120 percent minimum and 170 percent maximum of the character height.

703.2.4, 703.2.8
Visual Character Height and Line Spacing for Signs

703.2.4, 703.2.8 continues

CHANGE SIGNIFICANCE: The revisions to these sections came about because of the development of the VMS provisions of Section 703.7 and the realization that when signs are viewed at greater distances the standard does not adequately address how the characters should be dealt with and would literally require massive character heights.

The visual signage requirements were originally developed for viewing within a room or space where someone could move forward for a better view, such as moving closer to the train schedule in Grand Central station, rather than expecting that sign to be read from every possible location within the room. When dealing with a sports venue, it was realized that the signs had to be viewable from a person's seat without the person moving forward for better viewing. When developing the requirements for high-resolution VMS signage (see Section 703.7 discussion), the committee's original intent of the exception in 703.2.4 was to permit the height of the fixed characters and high-resolution VMS characters to use the same criteria for determining the minimum height of the characters. They quickly discovered that the required size was a problem when the score board or other sign was across the playing field or over the seating on the opposite side of the field. Without this exception, if Table 703.2.4 would be applied from each seat in an assembly space, the size of the sign would become extremely large and become the focus of the venue versus the actual purpose for the space.

In assembly seating, information is often presented in text form as well as audibly over the public address system. In large stadiums or venues this text information is viewed at great distances, potentially as much as 600 feet from some seats. While text information in these large assembly spaces is presented primarily in a VMS format, not all of it is always VMS. For example, a community sports facility may use a combination of fixed and VMS to minimize cost—characters that do not change very often, such as HOME and VISITOR on a scoreboard, may be fixed. Without an exception for general visual characters, these types of signs that comply with Table 703.2.4 and are viewed at great distances would result in signs that could be as large as the seating area.

The character height of 1 inch for every 30 feet in the exception was based upon the most recent *Manual on Uniform Traffic Control Devices* (MUTCD) requirements. Though these signs are not dealing with traffic signs, the MUTCD standard does address viewing distances for signs and is a standard that has been referenced by the A117.1 standard for years. See A117.1 Sections 105.1 and 105.2.1 for additional information related to the MUTCD standard.

An exception similar to that of Section 703.2.4 has been added into Section 703.2.8. The intent of this exception is to permit slightly reduced minimum line spacing for the fixed characters being viewed at larger distances. The 120 percent minimum character height spacing is similar to the criteria for VMS characters in the new Section 703.7.

CHANGE TYPE: Addition

CHANGE SUMMARY: An entire new section was added to address variable message signs (VMS). This section provides technical requirements for any VMS, such as those found in airports displaying flight arrival or gate information, and addresses both low- and high-resolution signs.

703.7
Variable Message Signs

2009 STANDARD:

Variable Message Signs (VMS).   Electronic signs that have a message with the capacity to change by means of scrolling, streaming, or paging across a background.

Variable Message Sign (VMS) Characters.   Characters of an electronic sign are composed of pixels in an array. High-resolution VMS characters have vertical pixel counts of 16 rows or greater. Low-resolution VMS characters have vertical pixel counts of 7 to 15 rows.

703.7 Variable Message Signs.

703.7.1 General.   High-resolution variable message sign (VMS) characters shall comply with Sections 703.2 and 703.7.12 through 703.7.14. Low-resolution variable message sign (VMS) characters shall comply with Section 703.7.

Exception: Theatrical performance related VMS signs, including but not limited to text and translation delivery systems, surtitles and subtitles, shall not be required to comply with Section 703.7.1.

703.7.2 Case.   Low-resolution VMS characters shall be uppercase.

Example 1

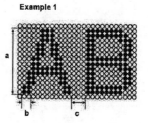

Example 2

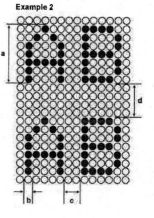

	Property	Example 1	Example 2
a	Character Height	14 Pixels	7 Pixels
b	Stroke Width	2 Pixels	1 Pixel
c	Character Spacing	3 Pixels	2 Pixels
d	Line Spacing		4 Pixels

703.7 continues

703.7 continued

703.7.3 Style. Low-resolution VMS characters shall be conventional in form, shall be san serif, and shall not be italic, oblique, script, highly decorative, or of other unusual forms.

703.7.4 Character Height. The uppercase letter "I" shall be used to determine the allowable height of all low-resolution VMS characters of a font. Viewing distance shall be measured as the horizontal distance between the character and an obstruction preventing further approach towards the sign. The uppercase letter "I" of the font shall have a minimum height complying with Table 703.7.4.

Exception: In assembly seating where the maximum viewing distance is 100 feet (30.5 m) or greater, the height of the uppercase "I" of low-resolution VMS fonts shall be permitted to be 1 inch (25 mm) for every 30 feet (9145 mm) of viewing distance, provided the character height is 8 inches (205 mm) minimum. Viewing distance shall be measured as the horizontal distance between the character and where someone is expected to view the sign.

TABLE 703.7.4   Low-Resolution VMS Character Height

Height above Floor to Baseline of Character	Horizontal Viewing Distance	Minimum Character Height
40 inches (1015 mm) to less than or equal to 70 inches (1780 mm)	Less than 10 feet (3050 mm)	2 inch (51 mm)
	10 feet (3050 mm) and greater	2 inch (51 mm), plus 1/5 inch (5.1 mm) per foot (305 mm) of viewing distance above 10 feet (3050 mm)
Greater than 70 inches (1780 mm) to less than or equal to 120 inches (3050 mm)	Less than 15 feet (4570 mm)	3 inches (75 mm)
	15 feet (4570 mm) and greater	3 inches (75 mm), plus 1/5 inch (5.1 mm) per foot (305 mm) of viewing distance above 15 feet (4570 mm)
Greater than 120 inches (3050 mm)	Less than 20 feet (6095 mm)	4 inches (100 mm)
	20 feet (6095 mm) and greater	4 inches (100 mm), plus 1/5 inch (5.1 mm) per foot (305 mm) of viewing distance above 20 feet (6095 mm)

TABLE 703.7.5   Pixel Count for Low-Resolution VMS Signage[1]

Character Height	Character Width Range	Stroke Width Range	Character Spacing Range
7	5–6	1	2
8	6–7	1–2	2–3
9	6–8	1–2	2–3
10	7–9	2	2–4
11	8–10	2	2–4
12	8–11	2	3–4
13	9–12	2–3	3–5
14	10–13	2–3	3–5
15	11–14	2–3	3–5

(1) Measured in pixels.

703.7.5 Character Width.   The uppercase letter "O" shall be used to determine the allowable width of all low-resolution VMS characters of a font. Low-resolution VMS characters shall comply with the pixel count for character width in Table 703.7.5.

703.7.6 Stroke Width.   The uppercase letter "I" shall be used to determine the allowable stroke width of all low-resolution VMS characters of a font. Low-resolution VMS characters shall comply with the pixel count for stroke width in Table 703.7.5.

703.7.7 Character Spacing.   Spacing shall be measured between the two closest points of adjacent low-resolution VMS characters within a message, excluding word spaces. Low-resolution VMS character spacing shall comply with the pixel count for character spacing in Table 703.7.5.

703.7.8 Line Spacing.   Low-resolution VMS characters shall comply with Section 703.2.8.

703.7.9 Height Above Floor.   Low-resolution VMS characters shall be 40 inches (1015 mm) minimum above the floor of the viewing position, measured to the baseline of the character. Heights of low-resolution variable message sign characters shall comply with Table 703.7.4, based on the size of the characters on the sign.

703.7.10 Finish.   The background of low-resolution VMS characters shall have a non-glare finish.

703.7.11 Contrast.   Low-resolution VMS characters shall be light characters on a dark background.

703.7.12 Protective Covering.   Where a protective layer is placed over VMS characters through which the VMS characters must be viewed, the protective covering shall have a non-glare finish.

703.7.13 Brightness.   The brightness of variable message signs in exterior locations shall automatically adjust in response to changes in ambient light levels.

703.7.14 Rate of Change.   Where a VMS message can be displayed in its entirety on a single screen, it shall be displayed on a single screen and shall remain motionless on the screen for a minimum 3 seconds or one second minimum for every 7 characters of the message including spaces whichever is longer.

CHANGE SIGNIFICANCE: VMS signs are used in a number of locations and may sometimes be of a type that might not traditionally be considered as a "sign." As an example, it is fairly common for airports to use television monitors or other similar devices to display flight arrival/departure or gate information. Although not signs in the traditional sense, these VMS signs provide information that is important and of use to the occupants so they may easily and effectively use the space.

703.7 continues

703.7 continued

VMS signs are classified as either high-resolution or low-resolution based upon the definition and the vertical pixel count. The high-resolution VMS signs are required to comply with the general visual character requirements of Section 703.2 as well as the VMS protective covering, brightness, and rate-of-change requirements in Section 703.7. The characters in low-resolution VMS signs are required to follow all of the requirements of Section 703.7. While it is becoming increasingly common to see high-resolution VMS, particularly LCD, plasma, and fiber-optic displays, that may be as legible as signs having conventional visual characters, low-resolution VMS, particularly signs that use either LED or electro-mechanical technology, will continue to be widely used.

Because of their legibility, the height of characters in low-resolution VMS signs must be greater than for conventional visual characters to result in equal legibility. Although many factors influence legibility, it is especially this need for greater character height, but also other design characteristics that facilitate legibility, that gives rise to the need for technical specifications for low-resolution VMS that are different than for signs having conventional visual characters. This character height difference can be seen by comparing the VMS requirements of Table 703.7.4 against the general visual character requirements of Table 703.2.4.

Section 703.7.2, similar to the requirements of Section 703.3, requires VMS characters to be in uppercase. A sign indicating "Chattanooga Choo Choo Track 29" using both upper- and lowercase letters, and having the required character height and so forth, would generally be less legible than providing the same message "CHATTANOOGA CHOO CHOO TRACK 29" in all uppercase letters. This legibility difference to readers is because of the size contrast between the upper- and lowercase characters and because the lowercase characters may appear as a lower resolution and be harder to read. For this reason, Section 703.7.2 requires that characters be consistent and shall all be uppercase.

The restriction on character styles in Section 703.7.3 is similar to the general visual character requirements of Section 703.2.3. Because of the low-resolution format, VMS characters with other than a fairly plain style would often not be legible or as easy to distinguish.

The character height, width, and spacing requirements of Sections 703.7.4 through 703.7.9 recognize that VMS are not legible from as great a distance as conventional signs having the same character height, and that, in part because of the matrix nature of VMS, different font geometries and spacing enhance legibility. Since character height is more closely related to legibility than any other sign characteristic, it makes the requirements of Section 703.7.4, Table 703.7.4, and Section 703.7.5 some of the most important aspects of these new requirements. The exception in the character height section recognizes that the general requirements for VMS were not ones that could readily be implemented in large assembly uses. There is, after all, a practical limitation to the overall size of these signs that can be used within the space. Therefore it is important to understand that the application of the exception is limited to assembly locations where viewing distances are 100 feet or greater and that the alternate height formula in the exception does not apply to other places in which large-format VMS may be used. Keep in mind that while a train station and a basketball arena might be the same size, the viewing distance is where the person can move to read the sign. In the train station, a person may walk forward for better viewing, thus possibly reducing the overall size of the sign. In the basketball arena, the viewpoint would be from every seat.

The intent of the rate of change provisions of Section 703.7.14 was simply to say that if it is possible to display an entire VMS message at one time, the message should be displayed in its entirety. Where the length of the message exceeds the capacity of the screen, the standard specifies a minimum length of time the message must be displayed before transitioning to the remaining portion of the message. The A117 committee did recognize that the reference to a "single screen" may create some confusion because some large VMS signs may be comprised of multiple smaller screens. Regardless of whether the VMS is a single screen or comprised of multiple screens, the intent remains the same: that if the message can be displayed on the sign "in its entirety" at a single time, that is the preferred option. Where it cannot be done at the same time, then the rate-of-change requirements apply.

708.4, 105.2.7

Telephone Entry Systems

CHANGE TYPE: Addition

CHANGE SUMMARY: Added a new reference standard and requires compliance with that standard for telephone entry systems. Provides a specific set of requirements and system performance for these systems. Will improve accessibility by making the systems more consistent and requiring them to meet the requirements of the added standard.

2009 STANDARD:

105.2.7 Performance Criteria for Accessible Communications Entry Systems. ANSI/DASMA 303-2006. (Door and Access Systems Manufacturers Association, 1300 Sumner Avenue, Cleveland, OH 44115-2851)

708.4 Telephone Entry Systems. Telephone entry systems shall comply with ANSI/DASMA 303 listed in Section 105.2.7.

CHANGE SIGNIFICANCE: The A117.1.1 standard previously did not address requirements for telephone entry systems. By not providing criteria for or addressing these systems, manufacturers, owners, and users were left without guidance for what is appropriate and a means to require the systems to meet a specific level of accessibility. The Door and Access Systems Manufacturers' Association (DASMA) standard solves this deficiency by providing appropriate references for all general requirements, while giving performance criteria on location, placement, visual user directions, audible user directions, volume level, call status, controls, and input devices. These types of systems are most commonly found at the front entrance to an apartment complex, but are increasingly used as part of the security for limited-access areas in facilities such as courthouses or high-tech facilities.

DASMA 303 is titled *Performance Criteria for Accessible Communications Entry Systems*. It was developed with the input of various users, producers, and others with a general interest in these types of entry systems. It provides a uniform means of evaluating the performance of accessible communication entry systems. The added text found in Section 708.4 of the A117.1 standard will provide the reference to require that the systems comply with the criteria of this newly developed DASMA standard.

The benefit for users of the A117.1 standard will be a uniform application of requirements toward such systems. The benefit toward users of the telephone entry systems themselves is an enhancement of systems performance and consistent ease of use. The benefit toward building owners is to enjoy the benefits telephone entry systems users will have, with minimal increase, if any, in construction cost.

The DASMA 300 standard is available for viewing and for downloading free of charge at the DASMA website, http://www.dasma.com/pubstandards.asp.

802
Assembly Areas

CHANGE TYPE: Modification

CHANGE SUMMARY: The assembly provisions have been modified to provide better guidance regarding the requirements and locations for wheelchair spaces within assembly areas. One significant aspect of these changes is creating a consistent terminology and distinguishing between a "wheelchair space" and "wheelchair space locations." (Dispersion of wheelchair space locations will be addressed in the next change.)

2009 STANDARD:

Wheelchair Space.   A space for a single wheelchair and its occupant.

Wheelchair Space Locations.   A space for a minimum of a single wheelchair and the associated companion seating. Wheelchair space locations can contain multiple wheelchair spaces and associated companion seating.

802 Assembly Areas

802.1 General. Wheelchair spaces and wheel chair space locations in assembly areas with spectator seating shall comply with Section 802. Team and player seating shall comply with Sections 802.2 through 802.6.

802.2 Floor Surfaces. The floor surface of wheelchair space locations shall have a slope not steeper than 1:48 and shall comply with Section 302.

802.3 Width. A single wheelchair space shall be 36 inches (915 mm) minimum in width. Where two adjacent wheelchair spaces are provided, each wheelchair space shall be 33 inches (840 mm) minimum in width.

802 continues

76

802.4 Depth. Where a wheelchair space ~~location~~ can be entered from the front or rear, the wheelchair space shall be 48 inches (1220 mm) minimum in depth. Where a wheelchair space ~~location~~ can only be entered from the side, the wheelchair space shall be 60 inches (1525 mm) minimum in depth.

802.5 Approach. The wheelchair space ~~location~~ shall adjoin an accessible route. The accessible route shall not overlap the wheelchair space ~~location~~.

802.5.1 Overlap. A wheelchair space ~~location~~ shall not overlap the required width of an aisle.

802.6 Integration of Wheelchair Space Locations. Wheelchair space locations shall be an integral part of any seating area.

802.7 Companion Seat. A companion seat, complying with Section 802.7, shall be provided beside each wheelchair space.

802.7.1 Companion Seat Type. The companion seat shall be ~~comparable~~ equivalent in size, ~~and~~ quality, comfort, and amenities ~~to assure equivalent comfort~~ to the seats ~~within the seating area adjacent~~ in the immediate area to the wheelchair space location. Companion seats shall be permitted to be moveable.

802.7.2 Companion Seat Alignment. In row seating, the companion seat shall be located to provide shoulder alignment with the wheelchair space occupant. The shoulder of the wheelchair space occupant ~~is considered to be 36 inches (915 mm) from the front of the wheelchair space.~~ shall be measured either 36 inches (915 mm) from the front or 12 inches (305 mm) from the rear of the wheelchair space. The floor surface for the companion seat shall be at the same elevation as the wheelchair space floor surface.

802.8 Designated Aisle Seats. Designated aisle seats shall comply with Section 802.8.

802.8.1 Armrests. Where armrests are provided on seating in the immediate area of designated aisle seats, folding or retractable armrests shall be provided on the aisle side of the designated aisle seat.

802.8.2 Identification. Each designated aisle seat shall be identified by ~~a sign or marker~~ the International Symbol of Accessibility.

802.9 Lines of Sight. Where spectators are expected to remain seated for purposes of viewing events, spectators in wheelchair space locations shall be provided with a line of sight in accordance with Section 802.9.1. Where spectators in front of the wheelchair space locations will be expected to stand at their seats for purposes of viewing events, spectators in wheelchair space locations shall be provided with a line of sight in accordance with Section 802.9.2.

802.9.1 Line of Sight over Seated Spectators. Where spectators are expected to remain seated during events, spectators seated in a wheelchair space ~~locations~~ shall be provided with lines of sight to the performance area or playing field comparable to that provided to <u>seated</u> spectators in closest proximity to the wheelchair space location. Where seating provides lines of sight over heads, spectators in wheelchair space locations shall be afforded lines of sight complying with Section 802.9.1.1. Where wheelchair space locations provide lines of sight over the shoulder and between heads, spectators in wheelchair space locations shall be afforded lines of sight complying with Section 802.9.1.2.

802.9.1.1 Lines of Sight over Heads. Spectators seated in<u> a</u> wheelchair space ~~locations~~ shall be afforded lines of sight over the heads of seated individuals in the first row in front of the wheelchair space location.

802.9.1.2 Lines of Sight between Heads. Spectators seated in <u>a</u> wheelchair space ~~locations~~ shall be afforded lines of sight over the shoulders and between the heads of seated individuals in the first row in front of the wheelchair space location.

802.9.2 Line of Sight over Standing Spectators. Wheelchair space<u>s</u> ~~locations~~ required to provide a line of sight over standing spectators shall comply with Section 802.9.2.

802.9.2.1 Distance from Adjacent Seating. The front of the wheelchair space <u>in a wheelchair space</u> location shall be 12 inches (305 mm) maximum from the back of the chair or bench in front.

802.9.2.2 ~~Elevation~~ <u>Height</u>. The ~~elevation~~ <u>height</u> of the ~~tread on which a~~ <u>floor surface at the</u> wheelchair space location ~~is located~~ shall comply with Table 802.9.2.2. <u>Interpolations shall be permitted for</u> riser heights ~~other than those provided, interpolations shall be permitted~~ <u>that are not listed in the table.</u>

CHANGE SIGNIFICANCE: While many of the changes in these sections will appear minor because they are simply clarifying whether a *wheelchair space* or a *wheelchair space location* is regulated by the provisions, this revision and the consistency of using the terms throughout Section 802 will help users determine what the standard is really looking for in each situation. These changes will help clarify when the requirements are for individual wheelchair spaces or groups of wheelchair spaces and the associated companion seat. In general, the technical requirements (depth, approach, need for a companion seat, and line of sight) will apply to a wheelchair space, while the dispersion requirements (horizontal, distance from event, and by type) will apply to wheelchair space locations. As stated earlier, the dispersion issues will be addressed in the change with Section 802.10.

An example of where the distinction is important and can affect the application of the provisions can be found in the approach requirements of Section 802.5. By deleting the word "location" in the first sentence of Section 802.5, the standard is clearly stating that each individual wheelchair space is required to adjoin an accessible route that provides access directly to the space. Contrast that requirement with the difference in

802 continues

802 continued

application if the provision applied to a "wheelchair space location." If applied to a "location," it would be permissible to provide one accessible route to the location even if there were four separate individual spaces within that location. Such a layout may mean that the individuals in three of the spaces may be required to move completely out of their space or the location in order to allow the last individual to access or leave his or her space. Conceptually, that would be similar to where the IBC permits the access to a wheelchair space in an area of refuge to pass through or be obstructed by an adjoining wheelchair space. However, even in that emergency situation the IBC permits no more than one adjoining wheelchair space as an obstruction from the accessible route.

The added requirement in Section 802.1 ensures that team and player seating require an integrated wheelchair space but do not require an adjacent companion seat, line of sight, dispersion, or a designated aisle seat.

The companion seating provisions of Section 802.7 will help to address several areas where there had previously been confusion or uncertainty. The type of seat to be used for the companion seat is to be essentially equivalent to that of the other seats near the wheelchair space location. This part of the changes in the companion seating provisions helps coordinate the standard with the ADA and ABA AG. The primary technical change in the companion seat requirements is found within Section 802.7.2. Where side access into a wheelchair space requires a 60-inch depth (Section 802.4), the previous standard's single option of measuring from the front of the space did not result in the shoulder-to-shoulder alignment that is required for the companion seat and the wheelchair space. The standard will use either the 36 inches measured from the front or the 12 inches measured from the back, depending on the location of the wheelchair space, so that it will align with the shoulders of the companion seat.

The remaining changes shown in the text help to clarify requirements and provide better guidance on the application of the standard. The changes in the line-of-sight sections clarify that the requirements are applicable to seated viewing arrangements from the wheelchair space and that the provisions apply to the *space* and not the *location*.

CHANGE TYPE: Modification

CHANGE SUMMARY: The dispersion of wheelchair space locations has been modified to eliminate some of the requirements that were difficult to apply to small venues, to coordinate with the ADA and ABA AG, and to help clarify the requirements. In addition, one aspect of these changes is creating a consistent terminology and distinguishing between a "wheelchair space" and "wheelchair space locations." (See the previous page covering Section 802 for additional related information.)

2009 STANDARD:

Wheelchair Space.   A space for a single wheelchair and its occupant.

Wheelchair Space Locations.   A space for a minimum of a single wheelchair and the associated companion seating. Wheelchair space locations can contain multiple wheelchair spaces and associated companion seating.

802.10 Wheelchair Space Dispersion.   The minimum number of wheelchair ~~spaces~~ locations shall be ~~dispersed to the minimum number of locations~~ in accordance with Table 802.10. Wheelchair space locations shall be dispersed in accordance with Sections 802.10.1, 802.10.2, and 802.10.3. In addition, ~~in spaces utilized primarily for viewing motion picture projection,~~ wheelchair space locations shall be dispersed in accordance with Section 802.10.4 in spaces utilized primarily for viewing motion picture projection. Once the required number of wheelchair space locations has been met, further dispersion is not required.

802.10.1 Horizontal Dispersion.   Wheelchair space locations shall be dispersed horizontally to provide viewing options. ~~Locations shall be separated by a minimum of 10 intervening seats.~~ Two wheelchair spaces shall be permitted to be located side-by-side.

802.10
Wheelchair Space Dispersion

802.10 continues

802.10 continued

Exception: Horizontal dispersion shall not be required in assembly areas with 300 or fewer seats if the wheelchair space locations are located within the 2nd and 3rd quartile of the row length. Intermediate aisles shall be included in determining the total row length. If the row length in the 2nd and 3rd quartile of the row is insufficient to accommodate the required number of companion seats and wheelchair spaces, the additional companion seats and wheelchair spaces shall be permitted to extend into the 1st and 4th quartile of the row. ~~In venues where wheelchair space locations are provided on only one side or on two opposite sides of the performance area or playing field, horizontal dispersion is not required where the locations are within the 2nd or 3rd quartile of the total row length. The wheelchair space locations and companion seats shall be permitted to overlap into the 1st and 4th quartile of the total row length if the 2nd and 3rd quartile of the row length does not provide the required length for the wheelchair space locations and companion seats. All intermediate aisles shall be included in determining the total row length.~~

TABLE 802.10 Wheelchair Space Location Dispersion

Total seating in Assembly Areas	Minimum required number of ~~dispersed~~ wheelchair space locations
Up to 150	1
151 to 500	2
501 to 1,000	3
1,001 to 5,000	3, plus 1 additional space for each 1,000 seats or portions thereof above 1,000
5,001 and over	7, plus 1 additional space for each 2,000 seats or portions thereof above 5,000

802.10.2 Dispersion for Variety of Distances from the Event. Wheelchair space locations shall be dispersed at a variety of distances from the event to provide viewing options. ~~Locations shall be separated by a minimum of five intervening rows.~~

Exceptions:

1. In bleachers, wheelchair space locations ~~shall not be required to be~~ provided only in rows ~~other than rows~~ at points of entry to bleacher seating shall be permitted.

2. ~~In spaces~~ Assembly areas utilized for viewing motion picture projections~~, assembly spaces~~ with 300 seats or less shall not be required to comply with Section 802.10.2.

3. ~~In spaces~~ Assembly areas with 300 seats or less other than those utilized for viewing motion picture projections~~, assembly spaces with 300 seats or less~~ shall not be required to comply with Section 802.10.2 ~~if the~~ where all wheelchair space locations are within the front 50 percent of the total rows.

802.10.3 Dispersion by Type. Where ~~there are~~ assembly seating ~~areas~~ has multiple distinct seating areas with amenities that differ from other distinct seating areas, ~~each having distinct services or amenities,~~ wheelchair space locations shall be provided within each distinct seating area.

802.10.4 Spaces Utilized Primarily for Viewing Motion Picture Projections.

In spaces utilized primarily for viewing motion picture projections, wheelchair space locations shall comply with Section 802.10.4.

802.10.4.1 Spaces with Seating on Risers.

Where tiered seating is provided, wheelchair space locations shall be integrated into the tiered seating area.

802.10.4.2 Distance from the Screen.

Wheelchair space locations shall be located within the rear ~~70~~ 60 percent of the seats provided.

CHANGE SIGNIFICANCE: For information related to the distinction of "wheelchair space" versus "wheelchair space location" see the previous page dealing with Section 802.

The horizontal dispersion provisions were modified by removing the requirement for the locations to be separated by a minimum of 10 seats. This previous requirement was eliminated primarily because it was difficult to comply with in small venues. Additional justification for its removal was that the provision did not guarantee that equivalent viewing angles were being provided, and this was not a specific requirement in the ADA and ABA AG. The exception has been revised so that it applies only to smaller assembly areas with 300 or fewer seats. Previously this exception could be used in any size assembly space and applied where the wheelchair space locations were "provided on only one side or on two opposite sides of the performance area or playing field." While the application of the exception is more limited, the methodology of placing the seats in the center portion of the row remains the same.

The dispersion of locations to provide a variety of distances from the event has been modified for reasons similar to the horizontal dispersion. That again is done because the previous provisions did not guarantee equivalent viewing angles, because of the difficulty in applying the requirements to small venues, and for consistency with the ADA and ABA AG. The only technical change is found in the base paragraph, where the previous requirement for the locations to be separated by a minimum of five intervening rows has been eliminated. This five-row separation was difficult if not impossible to apply in very small venues.

Because the 10 intervening seats and 5 intervening row requirements have been deleted, the standard will not provide the designer or inspector with a specifically required separation. Therefore this is an issue that should be discussed during the design stage to ensure that an agreement is reached early in the process so there is an adequate distribution. While the actual separation will depend on the size of the assembly space, it clearly is the intent of the standard that both horizontal dispersion and a variety of distances from the event are provided. As the number of wheelchair space locations increases, it is appropriate for even greater distribution of the locations.

The most significant change in the dispersion requirements can be found in Section 802.10.4.2 and is applicable to spaces utilized primarily for viewing projected motion pictures. This section was modified so that the wheelchair space locations were required to be in the rear 60 percent of the seats versus the rear 70 percent as was previously required. The 60 percent figure was selected for consistency with several settlement

802.10 continues

802.10 *continued*

agreements and consent orders that were worked out with the U.S. Department of Justice (DOJ) and theater owners. The 60 percent requirement was one of the choices that the DOJ allowed for determining the wheelchair space locations in existing movie theater auditoriums with less than 300 seats.

In the opinion of this author, this change to the 60 percent requirement was not adequately justified, was a misapplication of the DOJ settlements, and was not needed. First of all, it is important to realize that the DOJ agreements were worked out based on the fairly nebulous concept of providing equivalent lines of sight. While I agree with the concept of equal accommodation and the fairness this provides, there is no solid basis for establishing a specific percentage based strictly on the line-of-sight provisions. Because the ADA AG provisions do not provide specific requirements, what I may perceive to be equivalent will vary from what the next person deems acceptable. Using such vague concepts and details to establish a requirement makes for impossible design, application, and enforcement. Even at this point in time the DOJ has not finalized its rules to include the 60 percent requirement, nor has the U.S. Access Board included it into the ADA and ABA AG.

Second, these settlements and consent orders were based on trying to provide improved access to facilities that were not designed or constructed with any level of dispersion. By trying to retroactively provide dispersion, the options for a solution are and probably should be more limited. If a movie theater is designed and constructed using either the 2003 edition or this new edition of the A117.1 standard, then both the issues of lines of sight and dispersion are specifically addressed. With the standard requiring three different types of dispersion (horizontal, variety of distance, and by type) in very clear requirements, there is no need for the restrictive 60 percent option that made sense when no other dispersion was provided. Remember that neither the earlier ADA AG nor the new ADA and ABA AG contain specific dispersion requirements similar to those of the A117.1 standard. The 60 percent figure was developed simply as an opinion of what the line-of-sight provisions were intended to do and as one option to demonstrate compliance. Because the A117.1 standard contains other provisions on dispersion, the standard already addresses this issue, and the increased restriction of the 60 percent requirement was not needed.

The third issue that I feel affects this item is the fact that pushing the wheelchair space locations back farther into the seating has the potential to create additional problems related to the means of egress. Although this is a building code issue and not a requirement from the standard, it does affect the accessibility to the building and also the safety of people using these wheelchair spaces. As stated earlier, most of the theaters for which the DOJ settlements were developed were existing theaters that had already been constructed. When looking at the building code there is an exception that eliminates the need for an accessible means of egress in an existing building. Remember also that some of the theaters affected did not use the stadium seating design that is typical today. Therefore when dealing with these existing facilities, applying the 60 percent rule may be much easier because you do not need to worry about providing two accessible means of egress, and some of these theaters did not have the steep seating patterns. As you push the wheelchair space location back farther into the seating it becomes more difficult to provide two properly separated accessible means of egress. This then results in making a number of changes to the typical theater design and a greater amount of space being needed simply to satisfy the accessibility requirements. Therefore I do not see that the return on investment between the cost of these changes and a corresponding benefit or improvement in accessibility is substantial by simply pushing the seats 10 percent farther back in the space. Besides, seat selection and people's opinion of what location is better is a personal issue. While one person likes to sit near or at the back of the movie theater the next person likes to be closer to the front. Tightening the seating requirement to only the back 60 percent of the theater limits the options that some people may prefer.

Finally, and as covered briefly in an earlier paragraph, is the issue that the 60 percent figure accepted by the DOJ was *one* of several dispersion options that were deemed acceptable. The DOJ Settlement Agreements and Consent Orders related to movie theater dispersion provide designers and owners with several options. The 60 percent option is only one of those possible solutions, and that 60 percent requirement would not even be applicable if a designer selected one of the alternate methods of compliance. Therefore it seems unreasonable to me that despite the fact that dispersion is addressed within other sections of the A117.1 standard and that the DOJ also allows other means of compliance that the A117.1 standard would need to impose this more restrictive requirement in addition to or despite all of the other dispersion options. It is for these reasons that I feel the change to 60 percent was not adequately justified and was not a rational decision.

804

Kitchens and Kitchenettes in Common Use Spaces and Accessible Units

CHANGE TYPE: Modification

CHANGE SUMMARY: Numerous changes have been made for kitchens, including the elimination of a percentage of cabinet space for required accessible storage and clarification of the work surface requirements.

2009 STANDARD:

804 Kitchens and Kitchenettes

804.1 General. Accessible kitchens and kitchenettes shall comply with Section 804.

804.2 Clearance. Where a pass-through kitchen is provided, clearances shall comply with Section 804.2.1. Where a U-shaped kitchen is provided, clearances shall comply with Section 804.2.2.

Exception: Spaces that do not provide a cooktop or conventional range shall not be required to comply with Section 804.2 provided there is a 40 inch (1015 mm) minimum clearance between all opposing base cabinets, countertops, appliances, or walls within work areas.

804.3 Work Surface. At least one work surfaces shall comply be provided in accordance with Section 902.

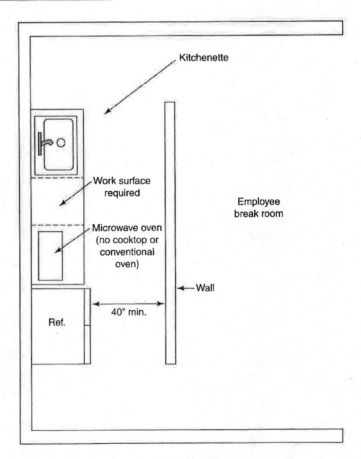

Exception: Spaces that do not provide a cooktop or conventional range shall not be required to provide an accessible work surface.

804.4 Sinks. ~~Sinks~~ The sink shall comply with Section 606.

804.5 Storage. ~~At least 50 percent of shelf space in cabinets shall comply with Section 905.~~

804.5 Appliances. Where provided, kitchen appliances shall comply with Section 804.5.

804.5.1 Clear Floor Space. A clear floor space complying with Section 305 shall be provided at each kitchen appliance. ~~Clear floor spaces are permitted to overlap.~~

CHANGE SIGNIFICANCE: Numerous changes affect the accessibility of a kitchen or kitchenette and the various appliances and elements within them. These changes are important because they apply not only to kitchens and kitchenettes within office spaces, churches, and so forth, but are also applicable to Accessible dwelling units because of the reference from Section 1002.12 to Section 804.

The additional text within Section 804.2 requires that the minimum clear space between opposing cabinets is required even where a cooktop or conventional oven is not provided. Because the Accessible dwelling units can be used in lieu of Type A and Type B dwelling units, it is appropriate that this requirement from Sections 1003.12.1.1 and 1004.12.1.1 be included here to avoid conflicts with the lesser levels of access required in the FHA and the ADA and ABA AG for dwelling units.

804 continues

804.5.4, 804.5.5, 1003.12.5.4, 1003.12.5.5, 1004.12.2.3, 1004.12.2.4

Cooktops and Ovens

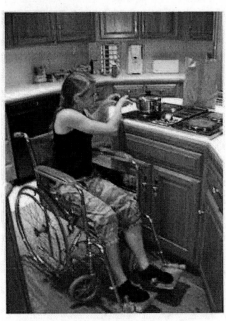

CHANGE TYPE: Modification

CHANGE SUMMARY: The revision works to coordinate the cooktop and oven requirements of Chapter 8 and those of the three dwelling unit types in Chapter 10. One primary change is that the range requirements are split out from the cooktop provisions.

2009 STANDARD:

804 Kitchens and Kitchenettes

804.5.4 ~~Range or~~ Cooktop. <u>Cooktops shall comply with Section 804.5.4.</u>

804.5.4.1 Approach. A clear floor space, positioned for a parallel or forward approach to the ~~space for a range or~~ cooktop, shall be provided.

804.5.4.2 Forward Approach. Where the clear floor space is positioned for a forward approach, knee and toe clearance complying with Section 306 shall be provided. ~~Where knee and toe space is provided,~~ <u>The</u> underside of the ~~range or~~ cooktop shall be insulated or otherwise configured to prevent burns, abrasions, or electrical shock.

804.5.4.3 Parallel Approach. <u>Where the clear floor space is positioned for a parallel approach, the clear floor space shall be centered on the appliance.</u>

804.5.4.4 Controls. The location of controls shall not require reaching across burners.

804.5.5 Oven. Ovens shall comply with Section 804.5.5.

804.5.5.1 Clear Floor Space. <u>A clear floor space complying with Section 305 shall be provided. The oven door in the open position shall not obstruct the clear floor space for the oven.</u>

804.5.5.4 Controls. ~~Ovens shall have controls on front panels~~. <u>The location of controls shall not require reaching across burners.</u>

1003 Type A Units

1003.12.5.4 ~~Range or~~ Cooktop. <u>Cooktops shall comply with Section 1003.12.5.4.</u>

1003.12.5.4.1 Approach. A clear floor space, positioned for a parallel or forward approach to the ~~space for a range or~~ cooktop, shall be provided.

1003.12.5.4.2 Forward Approach. Where the clear floor space is positioned for a forward approach, knee and toe clearance complying with Section 306 shall be provided. ~~Where knee and toe space is provided,~~ <u>The</u> underside of the ~~range or~~ cooktop shall be insulated or otherwise configured to protect from burns, abrasions, or electrical shock.

1003.12.5.4.3 Parallel Approach. Where the clear floor space is positioned for a parallel approach, the clear floor space shall be centered on the appliance.

1003.12.5.4.4 Controls. The location of controls shall not require reaching across burners.

1003.12.5.5 Oven. Ovens shall comply with Section 1003.12.5.5. Ovens shall have controls on front panels, on either side of the door.

1003.12.5.5.1 Clear Floor Space. A clear floor space shall be provided. The oven door in the open position shall not obstruct the clear floor space for the oven.

1003.12.5.5.2 Side-Hinged Door Ovens. Side-hinged door ovens shall have a countertop positioned adjacent to the latch side of the oven door.

1003.12.5.5.3 Bottom-Hinged Door Ovens. Bottom-hinged door ovens shall have a countertop positioned adjacent to one side of the door.

1003.12.5.5.4 Controls. The location of controls shall not require reaching across burners.

1004 Type B Units

1004.12.2.3 Cooktop. Cooktops shall comply with Section 1004.12.2.3.

1004.12.2.3.1 Approach. A clear floor space, positioned for a parallel or forward approach to the cooktop, shall be provided. ~~The centerline of the clear floor space shall align with the centerline of the cooktop.~~

1004.12.2.3.2 Forward Approach. Where the clear floor space is positioned for a forward approach, knee and toe clearance complying with Section 306 shall be provided. ~~Where knee and toe space is provided,~~ The underside of the ~~range or~~ cooktop shall be insulated or otherwise configured to prevent burns, abrasions, or electrical shock.

1004.12.2.3.3 Parallel Approach. Where the clear floor space is positioned for a parallel approach, the clear floor space shall be centered on the appliance.

1004.12.2.4 Oven. A clear floor space, positioned for a parallel or forward approach ~~to the oven,~~ adjacent to the oven shall be provided. The oven door in the open position shall not obstruct the clear floor space for the oven.

CHANGE SIGNIFICANCE: This change includes revisions in Chapter 8 that apply to the general kitchen requirements and those of the Accessible units, with similar changes being made in Section 1003 for the Type A units and Section 1004 for Type B units. The previous provisions dealing

804.5.4, 804.5.5, 1003.12.5.4, 1003.12.5.5, 1004.12.2.3, 1004.12.2.4 continues

804.5.4, 804.5.5, 1003.12.5.4,
1003.12.5.5, 1004.12.2.3,
1004.12.2.4 continued

with ranges and cooktops have been revised so that the requirements apply to separate cooktops or to the cooktop portion of a range. This change in application results from the reference to ranges being deleted from Sections 804.5.4 and 1003.12.5.4. A range is now addressed as a cooktop and an oven combined. Therefore Sections 804.5.4, 1003.12.5.4, and 1004.12.2.3 will apply to cook tops and Sections 804.5.5, 1003.12.5.5, and 1004.12.2.4 will apply to ovens. Both sets of requirements will apply to ranges. The requirements for the cook tops have been broken into smaller sections, with the primary technical change being that within the general kitchen requirements of Chapter 8 and the Type A dwelling unit requirements of Section 1003.12.5.4.3, the parallel-approach clear floor space is required to be centered on the appliance it serves. The requirement for the centering of the clear floor space helps match a provision that has existed within the Type B units. When using a side approach it is important that the clear floor space and the appliance are closely aligned to allow access to all burners.

In regard to ovens, the changes include a new section that requires the clear floor space that serves the appliance to be located so the space is not obstructed when the door to the oven is opened. Therefore the location of the clear floor space needs to consider how the oven door operates and be located so that the door can be opened and access to the oven can be maintained. The clear floor space can use the knee and toe clearances available when the door is open. In other than the Type B units, the control requirements have been revised so that they are no longer required to be at the front of the appliance but could be located in other locations such as to the side of the burners. Because appliance controls are not regulated in Type B units, a similar change was not needed for the Type B section.

Keep in mind that in general kitchens and Accessible unit kitchens, the accessible working surface must be adjacent to the oven, and in a Type A unit a counter must be available adjacent to the oven.

Users may want to pay special attention to Section 1003.12.5.5 and the fact it still contains the sentence "Ovens shall have controls on front panels, on either side of the door." It is my opinion that this sentence should have been eliminated based on the intent of the actions the A117 committee took when they approved this code change (Change proposal #267). It seems clear to me that based on the revisions in Sections 804.5.5 and 805.5.5.4, and the reason statements given when approving the proposal, that a similar coordinating change was appropriate for the Type A units. However, due to the fact that the code text was shown incorrectly in the published code change proposal, this sentence had to be kept even though the new 1003.12.5.5.4 was added. The fact that Sections 1003.12.5.5 and 1003.12.5.5.4 both exist will not create a conflict but could create confusion, especially because they differ from the requirements for kitchens in Chapter 8 and in the Accessible units. With this text remaining, a wall oven with controls located above or below the oven door is not permitted in Type A units. In other unit types and in public kitchen areas, the only requirement is that the controls be located so a user does not need to reach across the burners.

CHANGE TYPE: Modification

CHANGE SUMMARY: Provision requires detectable warnings any time a track crosses a pedestrian way. Previously the requirement only applied to circulation paths serving a boarding platform.

805.10
Track Crossings

2009 STANDARD:

805.10 Track Crossings.   Where a circulation path ~~serving boarding platforms~~ crosses tracks, it shall comply with Section 402 and shall have a detectable warning 24 inches (610 mm) in depth complying with Section 705 extending the full width of the circulation path. The detectable warning surface shall be located so that the edge nearest the rail crossing is 6 foot (1.8 m) minimum and 15 foot (4.6 m) maximum from the centerline of the nearest rail.

Exception: Openings for wheel flanges shall be permitted to be 2½ inches (64 mm) maximum.

CHANGE SIGNIFICANCE: Crossing a track presents the same hazard whether or not it serves a boarding platform. As previously written the requirement only applied to a circulation path "serving boarding platforms." This revised text will require that any pedestrian circulation path that crosses a set of tracks be provided with detectable warnings. Therefore the scope of this section has expanded to regulate the circulation path regardless of whether it serves the boarding platform or not. An example of where this could occur would be in an amusement park where a train may cross the pedestrian walkway or if there was an accessible route on a site where a sidewalk may cross over a track. One item to note is because this section is not tied to the transportation platform, the jurisdiction would need to provide scoping provisions to specify where these technical requirements apply.

805.10 continues

807 continued

Exception: ~~Levels of jury boxes not required to be accessible are not required to comply with Section 807.2.~~

807.3 Clear Floor Space. Within the defined area of each jury box and witness stand, a clear floor space complying with Section 305 shall be provided.

Exception: In alterations, wheelchair spaces are not required to be located within the defined area of raised jury boxes or witness stands and shall be permitted to be located outside these spaces where ramps or platform lifts restrict or project into the means of egress required by the administrative authority.

807.4 ~~Judges'~~ Benches and Courtroom Stations. Judges' benches, clerks' stations, bailiffs' stations, deputy clerks' stations, court reporters' stations, and litigants' and counsel stations shall comply with Section 902.

807.5 Gallery Seating. ~~Gallery seating shall comply with Section 802.~~

CHANGE SIGNIFICANCE: One of the primary purposes for these changes was to coordinate with Section 808 of the ADA and ABA AG and allowances in the IBC Section 1108.4, but they will result in the A117.1 being a bit more restrictive than previously when dealing with the turning space requirements of Section 807.2. The primary changes in the courtroom requirements are found in Section 807.2 and in the new Section 807.5. The revised title in Section 807.4 is strictly an editorial issue since a judge's bench is a courtroom station and it is specifically included along with the others in the list within the section.

The turning space requirements of Section 807.2 address four separate issues. The changes all help to ensure that the spaces are accessible as they are intended to be, with the exception that the vertical portion of an accessible route is not required. The four changes and their purpose are as follows:

1. Deleting the text "and accessed by ramps or platform lifts": This previous text provided a second trigger for when raised or depressed areas required a turning space. The turning space is now required for any raised or depressed area regardless of whether a ramp or lift is provided. Since court employee work stations can be adaptable, this ensures the turning space is available when constructed and then only requires the installation of a lift or ramp as a part of the adaptation.

2. Deleting the text ". . . platform lifts with entry ramps": This previous text would literally only require the turning space when the platform lift had an entry ramp. This change helps coordinate with the platform lift door and gate requirements found in Section 410.2.1. The platform lift requirements allow a manual door in several situations and could also require that the user be properly located to maneuver onto the lift. The turning space allows the user to maneuver to access the lift and to operate the doors even where they do not include an entry ramp. As an

example, a lift complying with Section 410.2.1 Exception 1 could require the persons on the raised area to reach behind them to operate a door on the lift if they could not turn to approach the lift in a forward direction.

3. Deleting the text "an unobstructed": The previous requirement for *an unobstructed* turning space led some people to interpret that the space was not permitted to include knee and toe clearances as permitted by Section 304.3. The deletion of this text should help to eliminate the confusion and clarify that the knee and toe clearances under the work surfaces (i.e., judge's bench or clerk's counter) are permitted in the turning space.

4. Addition of the exception concerning jury boxes: Generally, jury boxes only require one of the seating levels to be accessible due to the fact that only one accessible clear floor space is required within the jury box. This exception eliminates the requirement that would otherwise be imposed on "*each* area that is raised or depressed." It is reasonable in a jury box to accept access to only one level. Jury boxes often have two or more tiers to provide a line of sight for the jurors.

Section 807.5 provides a reference to Section 802 and clarifies that the gallery seating in the courtroom is dealt with the same as any other assembly seating. This includes the requirements for wheelchair spaces, companion seating, line of sight, dispersion, and designated aisle seats. See the pages dealing with Section 802 for information on changes within that section.

902

Clear Floor Space for Dining and Work Surfaces

Courtesy Populous

CHANGE TYPE: Modification

CHANGE SUMMARY: Two new exceptions are provided to eliminate the requirement for knee and toe space beneath small drink counters or furnishings where the element is not intended to function as a table for forward approach.

2009 STANDARD:

902 Dining Surfaces and Work Surfaces

902.1 General. Accessible dining surfaces and work surfaces shall comply with Section 902.

Exception: Dining surfaces and work surfaces primarily for children's use shall be permitted to comply with Section 902.5.

902.2 Clear Floor Space. A clear floor space complying with Section 305, positioned for a forward approach, shall be provided. Knee and toe clearance complying with Section 306 shall be provided.

Exceptions:

1. At drink surfaces 12 inches (305 mm) or less in depth, knee and toe space shall not be required to extend beneath the surface beyond the depth of the drink surface provided.
2. Dining surfaces that are 15 inches (380 mm) minimum and 24 inches (610 mm) maximum in height are permitted to have a clear floor space complying with Section 305 positioned for a parallel approach.

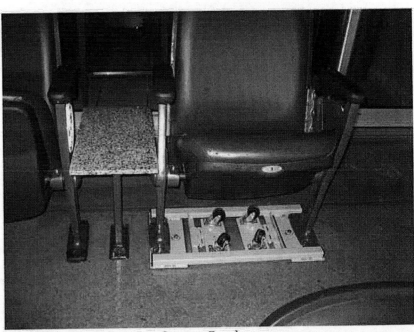

Courtesy Populous

902.3 Exposed Surfaces. There shall be no sharp or abrasive surfaces under the exposed portions of dining surfaces and work surfaces.

CHANGE SIGNIFICANCE: The two new exceptions provide relief from the requirement for knee and toe space to be provided beneath an element. These types of elements are not really designed or intended to function as a table, dining, or work surface but often can be viewed that way. As an example of the type of elements covered by Exception 1, consider a small countertop or ledge that is placed along the wall in a standing area of a bar or club. These drink ledges are often only a few inches in depth and are used simply as a place to set a drink or a plate of appetizers down. They are not intended to be used in a forward approach where someone would need to be able to get their knees beneath the element. If the knee and toe clearance were to be required beneath the element, then based on the requirements of Section 306, the minimum depth of the ledge would need to be at least 17 inches. Increasing the depth of the element would only increase the likelihood that it would be used by a forward approach; therefore the standard imposes a 12-inch limit to help distinguish when the exception applies.

Another problem that would arise by requiring the knee and toe clearance is that the clearance beneath the element would be required to be at least 27 inches above the floor. In situations where the drink ledge is attached to a guard at the front of an assembly seating area, this would mean that the height of the drink ledge would be required to be raised above the 26-inch guard height that is allowed by the building code. The IBC allows this lower guard height so the view from the first row of seats is not obstructed.

The second exception can be used for items such as the low side tables that are often found adjacent to lounge chairs in waiting areas of clubs or in the lobby of a hotel. Many times these tables are used as a location to place drinks and snacks but are not truly used as a dining surface. By requiring a parallel-approach clear floor space and also limiting the height, this exception provides a reasonable degree of accessibility while providing guidance for an accessible side table or possibly maintaining sight lines when they are used in an assembly seating area.

The addition of Section 902.3 addresses a requirement not previously covered in the standard. This requirement is conceptually similar to sections dealing with elements beneath cooktops, sinks, or lavatories and is intended to prevent harm to users who place their legs beneath an element. Regardless of the type of element, it is reasonable that there not be any sharp or abrasive surfaces exposed beneath it. Adding this text here extends this requirement to dining and work surfaces to specifically address this danger.

903

Benches in Locker Rooms, Fitting Rooms, and Dressing Rooms

CHANGE TYPE: Modification and addition

CHANGE SUMMARY: The clear floor space to access the bench is no longer required to be at the end of the bench. A parallel approach to the front of the bench is now permitted. A new exception that addresses the height of benches for children has been added.

2009 STANDARD:

903 Benches

903.1 General. Accessible benches shall comply with Section 903.

903.2 Clear Floor Space. A clear floor space complying with Section 305, positioned for parallel approach to ~~an end of~~ the bench seat, shall be provided.

903.3 Size. Benches shall have seats 42 inches (1065 mm) minimum in length, and 20 inches (510 mm) minimum and 24 inches (610 mm) maximum in depth.

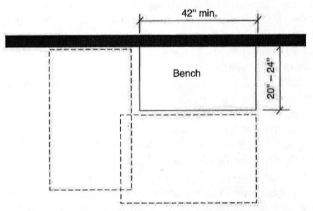

(a) Bench size and options for clear floor space

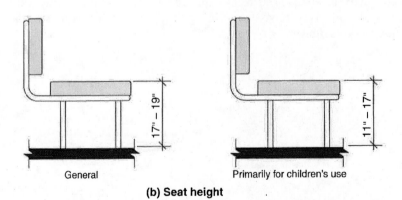

(b) Seat height

903.4 Back Support.   The bench shall provide for back support or shall be affixed to a wall. Back support shall be 42 inches (1065 mm) minimum in length and shall extend from a point 2 inches (51 mm) maximum above the seat surface to a point 18 inches (455 mm) minimum above the seat surface. Back support shall be 2-1/2 inches (64 mm) maximum from the rear edge of the seat measured horizontally.

903.5 Height.   The top of the bench seat shall be 17 inches (430 mm) minimum and 19 inches (485 mm) maximum above the floor, measured to the top of the seat.

Exception: Benches primarily for children's use shall be permitted to be 11 inches (280 mm) minimum and 17 inches (430 mm) maximum above the floor, measured to the top of the seat.

CHANGE SIGNIFICANCE: A bench is required within an accessible dressing room, fitting room, or locker room. Previously the standard required that the clear floor space to access the bench be located to provide a parallel approach at the end of the bench seat. With this change the clear floor space is allowed to be located either at the end or along the front edge of the bench. This revision will allow designers an option regarding where to locate the clear floor space in relation to the bench. The committee discussion indicated that the transfer can be made from any clear floor space adjacent to the bench and that the space does not have to be at the end. This revision to Section 903.2 provides adequate access and can save space.

Requiring the clear floor space at the end of the bench can unnecessarily drive the size and the layout of the room. This change may help reduce the size required for some dressing rooms since the 42-inch bench length plus a minimum 30-inch clear floor space at the end of the bench established a specific length for one side of the space that was also affected by the requirement of Section 803.2 for a turning space. Requiring the clear floor space to be positioned at the end of the bench was overly restrictive and provided little, if any, benefit compared to a parallel approach to the front of the bench.

If someone did position his or her chair to back in at the end of the bench, the wheels of the wheelchair would bump into the supporting wall and not allow a good side transfer position similar to that found at water closets or transfer showers. Testimony at the committee meeting indicated that the existing type of transfer is a difficult one to make and that most users typically do not back into the space but instead go in toes first or stay in front of the bench. Many wheelchair users will prefer the clear floor space positioned along the front edge of the bench.

As currently written, the standard does not prohibit an armrest or other barrier at the end of the bench. Obviously, the standard does intend for the person in the wheelchair to be able to transfer to the bench from the area of the clear floor space, but the text does not mention that obstructions should not limit the access onto the bench.

A new exception in Section 903.5 will provide a specific height limitation for benches intended primarily for the use of children. These benches must be located at a lower height than the height required for adults. This revision is a part of an effort to provide more detailed information within the standard for designs that are intended for children.

Chapter 10

Grab Bar and Shower Seat Reinforcement in Dwelling Units and Sleeping Units

CHANGE TYPE: Clarification

CHANGE SUMMARY: Many of the exceptions and provisions that apply specifically to blocking for grab bars and shower seats in residential dwelling and sleeping units have been relocated to Chapter 10 so that the dwelling unit and sleeping unit requirements are together in one location. In most situations this is just a relocation of the requirements.

2009 STANDARD:

604.5 Grab Bars. Grab bars for water closets shall comply with Section 609 and shall be provided in accordance with Sections 604.5.1 and 604.5.2. Grab bars shall be provided on the rear wall and on the side wall closest to the water closet.

Exceptions:

1. Grab bars are not required to be installed in a toilet room for a single occupant, accessed only through a private office and not for common use or public use, provided reinforcement has been installed in walls and located so as to permit the installation of grab bars complying with Section 604.5.

2. In detention or correction facilities, grab bars are not required to be installed in housing or holding cells or rooms that are specially designed without protrusions for purposes of suicide prevention.

3. ~~In Type A units, grab bars are not required to be installed where reinforcement complying with Section 1003.11.4 is installed for the future installation of grab bars.~~

4. ~~In Type B units located in institutional facilities and assisted living facilities, two swing-up grab bars shall be permitted to be installed in lieu of the rear wall and side wall grab bars. Swing-up grab bars shall comply with Sections 604.5.3 and 609.~~

5. ~~In a Type B unit, where fixtures are located on both sides of the water closet, a swing-up grab bar complying with Sections 604.5.3 and 609 shall be permitted. The swing-up grab bar shall be installed on the side of the water closet with the 18-inch (455 mm) clearance required by Section 1004.11.3.1.2.~~

604.5.1 Fixed Side-Wall Grab Bars. Fixed side-wall grab bars shall be 42 inches (1065 mm) minimum in length, located 12 inches (305 mm) maximum from the rear wall and extending 54 inches (1370 mm) minimum from the rear wall. In addition, a vertical grab bar 18 inches (455 mm) minimum in length shall be mounted with the bottom of the bar located ~~between~~ 39 inches (990 mm) <u>minimum</u> and 41 inches (1040 mm) <u>maximum</u> above the floor, and with the centerline of the bar located ~~between~~ 39 inches (990 mm) <u>minimum</u> and 41 inches (1040 mm) <u>maximum</u> from the rear wall.

Exceptions: <u>The vertical grab bar at water closets primarily for children's use shall comply with Section 609.4.2.</u>

  ~~1. In Type A and Type B units, the vertical grab bar component is not required.~~
  ~~2. In a Type B unit, when a side wall is not available for a 42-inch (1065 mm) grab bar, the sidewall grab bar shall be permitted to be 18 inches (455 mm) minimum in length, located 12 inches (305 mm) maximum from the rear wall and extending 30 inches (760 mm) minimum from the rear wall.~~

604.5.2 Rear-Wall Grab Bars. The rear-wall grab bar shall be 36 inches (915 mm) minimum in length, and extend from the centerline of the water closet 12 inches (305 mm) minimum on the side closest to the wall, and 24 inches (610 mm) minimum on the transfer side.

Chapter 10 continues

Chapter 10 continued   **Exceptions:**

1. The rear grab bar shall be permitted to be 24 inches (610 mm) minimum in length, centered on the water closet, where wall space does not permit a grab bar 36 inches (915 mm) minimum in length due to the location of a recessed fixture adjacent to the water closet.

2. ~~In a Type A or Type B unit, the rear grab bar shall be permitted to be 24 inches (610 mm) minimum in length, centered on the water closet, where wall space does not permit a grab bar 36 inches (915 mm) minimum in length.~~

~~3~~ 2. Where an administrative authority requires flush controls for flush valves to be located in a position that conflicts with the location of the rear grab bar, that grab bar shall be permitted to be split or shifted to the open side of the toilet area.

604.5.3 Swing-up Grab Bars. ~~Where swing-up grab bars are installed, a clearance of 18 inches (455 mm) minimum from the centerline of the water closet to any side wall or obstruction shall be provided. A swing-up grab bar shall be installed with the centerline of the grab bar 15¾ inches (400 mm) from the centerline of the water closet. Swing-up grab bars shall be 28 inches (710 mm) minimum in length, measured from the wall to the end of the horizontal portion of the grab bar.~~

1003 Type A Units

1003.11 Toilet and Bathing Facilities. At least one toilet and bathing facility shall comply with Section 1003.11.2. All toilet and bathing facilities shall comply with Section 1003.11.1.

1003.11.1 ~~1003.11.4~~ **Grab Bar and Shower Seat Reinforcement.** Reinforcement shall be provided for the future installation of grab bars complying with Section 604.5 at water closets, grab bars complying with Section 607.4 at bathtubs, and for grab bars and shower seats complying with Sections 608.3, 608.2.1.3, 608.2.2.3, and 608.2.3.2 at shower compartments. ~~and shower seats at water closets, bathtubs, and shower compartments. Where walls are located to permit the installation of grab bars and seats complying with Sections 604.5, 607.4, 608.3 and 608.4, reinforcement shall be provided for the future installation of grab bars and seats meeting those requirements.~~

Exceptions:

1. At fixtures not required to comply with Section 1003.11.2, reinforcement in accordance with Section 1004.11.1 shall be permitted.

2. Reinforcement is not required in a room containing only a lavatory and a water closet, provided the room does not contain the only lavatory or water closet on the accessible level of the dwelling unit.

3. Reinforcement for the water closet side-wall vertical grab bar component required by Section 604.5 is not required.

4. Where the lavatory overlaps the water closet clearance in accordance with the exception to Section 1003.11.2.4.4 reinforcement at the water closet rear wall for a 24 inch (610 mm) minimum-length grab bar, centered on the water closet, shall be provided.

1004 Type B Units

1004.11 Toilet and Bathing Facilities. Toilet and bathing fixtures shall comply with Section 1004.11.

Exception: Fixtures on levels not required to be accessible.

~~1004.11.2~~ 1004.11.1 Grab Bar and Shower Seat Reinforcement. Reinforcement shall be provided for the future installation of grab bars and shower seats at water closets, bathtubs, and shower compartments. Where walls are located to permit the installation of grab bars and seats complying with ~~Sections~~ 604.5 at water closets, grab bars complying with Section 607.4 at bathtubs, and for grab bars and shower seats complying with Sections, 608.3, 608.2.1.3, 608.2.2.3, and 608.2.3.2 at shower compartments, reinforcement shall be provided for the future installation of grab bars and seats ~~meeting~~ complying with those requirements.

Exceptions:

1. ~~Reinforcement is not required~~ In a room containing only a lavatory and a water closet, reinforcement is not required provided the room does not contain the only lavatory or water closet on the accessible level of the unit.

2. At water closets reinforcement for the side-wall vertical grab bar component required by Section 604.5 is not required.

3. At water closets where wall space will not permit a grab bar complying with Section 604.5.2, reinforcement for a rear-wall grab bar 24 inches (610 mm) minimum in length centered on the water closet shall be provided.

4. At water closets where a side wall is not available for a 42-inch (1065-mm) grab bar complying with Section 604.5.1, reinforcement for a side-wall grab bar, 24 inches (610 mm) minimum in length, located 12 inches (305 mm) maximum from the rear wall, shall be provided.

5. At water closets where a side wall is not available for a 42-inch (1065-mm) grab bar complying with Section 604.5.1 reinforcement for a swing-up grab bar complying with Section 1004.11.1.1 shall be permitted.

6. At water closets where a side wall is not available for a 42-inch (1065-mm) grab bar complying with Section 604.5.1 reinforcement for two swing-up grab bars complying with Section 1004.11.1.1 shall be permitted to be installed in lieu of reinforcement for rear-wall and side-wall grab bars.

7. In shower compartments larger than 36 inches (915 mm) in width and 36 inches (915 mm) in depth, reinforcement for a shower seat is not required.

Chapter 10 continues

Chapter 10 continued

1004.11.1 Swing-up Grab Bars. A clearance of 18 inches (455 mm) minimum from the centerline of the water closet to any side wall or obstruction shall be provided where reinforcement for swing-up grab bars is provided. When the approach to the water closet is from the side, the 18 inches (455 mm) minimum shall be on the side opposite the direction of approach. Reinforcement shall accommodate a swing-up grab bar centered 15-3/4 inches (400 mm) from the centerline of the water closet and 28 inches (710 mm) minimum in length, measured from the wall to the end of the horizontal portion of the grab bar. Reinforcement shall accommodate a swing-up grab bar with a height in the down position of 33 inches (840 mm) minimum and 36 inches (915 mm) maximum. Reinforcement shall be adequate to resist forces in accordance with Section 609.8.

Exception: Where a water closet is positioned with a wall to the rear and to one side, the centerline of the water closet shall be 16 inches (405 mm) minimum and 18 inches (455 mm) maximum from the sidewall.

CHANGE SIGNIFICANCE: The A117 committee made an effort to relocate many of the general code requirements and exceptions that apply specifically to dwelling and sleeping units from the general chapters to Chapter 10. This type of relocation of the requirements occurs for elements such as reinforcement for grab bars and shower seats. Moving the requirements to Chapter 10 allows the standard to have the specific provisions for dwelling and sleeping units located where they will be more apparent. This will help eliminate some of the confusion that can occur when an exception is shown in a section in Chapter 6 but it is only applicable to a dwelling and sleeping unit.

Although some of the new text may result in minor changes to the requirements, this effort to relocate the provisions will not typically affect the application of the standard. It is important that users are aware of this change so they do not assume the provision has been deleted when they look in the general chapters and see a dwelling unit provision from the 2003 standard is no longer at the same location. At the same time, when looking at the requirements of Chapter 10 in the new standard and seeing what appears to be a new requirement, they must realize that it may be a provision that has been moved and not an actual change in the requirements.

CHANGE TYPE: Modification

CHANGE SUMMARY: The requirements for the accessible route's location and entry into the dwelling or sleeping unit have been revised and coordinated for the three types of units.

2009 STANDARD:

1002 Accessible Units

1002.1 General.   Accessible units shall comply with Section 1002.

1002.2 Primary Entrance.   The accessible primary entrance shall be on an accessible route from public and common areas. The primary entrance shall not be to a bedroom <u>unless it is the only entrance.</u>

1002.3 Accessible Route.   Accessible routes within Accessible units shall comply with Section 1002.3. ~~Exterior spaces less than 30 inches (760 mm) in depth or width shall comply with Sections 1002.3.1, 1002.3.3, 302, and 303.~~

1002.3.1 Location.   At least one accessible route shall connect all spaces and elements that are a part of the unit. <u>Accessible routes shall coincide with or be located in the same area as a general circulation path.</u> ~~Where only one accessible route is provided, it shall not pass through bathrooms and toilet rooms, closets, or similar spaces.~~

Exception: An accessible route is not required to unfinished attics and unfinished basements that are part of the unit.

1003 Type A Units

1003.1 General.   Type A units shall comply with Section 1003.

1002.3, 1003.3, 1004.3

Accessible Route and Entrance Requirements for Accessible, Type A, and Type B Units

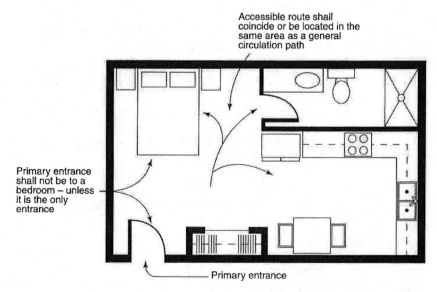

Accessible route shall coincide or be located in the same area as a general circulation path

Primary entrance shall not be to a bedroom – unless it is the only entrance

Primary entrance

1002.3, 1003.3, 1004.3 continues

1002.3, 1003.3, 1004.3 continued

1003.2 Primary Entrance. The accessible primary entrance shall be on an accessible route from public and common areas. The primary entrance shall not be to a bedroom unless it is the only entrance.

1003.3 Accessible Route. Accessible routes within Type A units shall comply with Section 1003.3. ~~Exterior spaces less than 30 inches (760 mm) in depth or width shall comply with Sections 1003.3.1, 1003.3.3, 302, and 303.~~

1003.3.1 Location. At least one accessible route shall connect all spaces and elements that are a part of the unit. Accessible routes shall coincide with or be located in the same area as a general circulation path. ~~Where only one accessible route is provided, it shall not pass through bathrooms and toilet rooms, closets, or similar spaces.~~

Exception: An accessible route is not required to unfinished attics and unfinished basements that are part of the unit.

1004 Type B Units

1004.1 General. Type B units shall comply with Section 1004.

1004.2 Primary Entrance. The accessible primary entrance shall be on an accessible route from public and common areas. The primary entrance shall not be to a bedroom unless it is the only entrance.

1004.3 Accessible Route. Accessible routes within Type B units shall comply with Section 1004.3.

1004.3.1 Location. At least one accessible route shall connect all spaces and elements that are a part of the unit. Accessible routes shall coincide with or be located in the same area as a general circulation path. ~~Where only one accessible route is provided, it shall not pass through bathrooms and toilet rooms, closets, or similar spaces.~~

Exceptions:

1. An accessible route is not required to unfinished attics and unfinished basements that are part of the unit.
2. One of the following is not required to be on an accessible route:
   2.1 A raised floor area in a portion of a living, dining, or sleeping room; or
   2.2 A sunken floor area in a portion of a living, dining, or sleeping room; or
   2.3 A mezzanine that does not have plumbing fixtures or an enclosed habitable space.

CHANGE SIGNIFICANCE: The primary entrance requirements of Sections 1002.2, 1003.2, and 1004.2 are intended to ensure that the primary accessible entrance is the same entrance that is used by persons without disabilities to enter their dwelling units and not a secondary door. The added text regarding situations where an entry through a bedroom "is the only entrance" is intended to resolve the uncertainty that occurs where the unit is either an efficiency unit or a sleeping unit in a hotel, dormitory, assisted living facility, and so forth. Without this added text the accessible entrance could not pass through the "bedroom" even where the entire unit is the bedroom/sleeping area.

The location requirements of Section 1002.3.1, 1003.3.1, and 1004.3.1 have been modified to create a general requirement that the accessible route coincide with the general circulation path. Adding the provision as a general requirement allows the text to address a variety of conditions and permits the elimination of the laundry list of spaces and the difficult-to-determine "similar spaces" from the last sentence of the paragraph. As long as the accessible route within the unit is the same as the route used by everyone else, it should not make any difference which rooms or spaces it goes through.

A new exception 1 is added to the Type B unit requirements of Section 1004.3.1. This added exception addresses unfinished attics and basements and coordinates with existing language from the Accessible and Type A units. If this exception is available for both an Accessible and Type A unit, then it is appropriate that the Type B units that are considered as a lower level of accessibility also include the exception. Adding this language is also consistent with the multistory unit exception that is found in the FHA and also in the IBC. When discussing the inclusion of this exception, the A117 committee did state that it believed the doors to these spaces were still a "user passage doorway" (see Section 1004.5.2) and that the door width and size requirements were still regulated. Therefore the committee's intent was that this exception does not exempt doors to these unfinished spaces from complying with Section 1004.5.

The deletion of the text from Sections 1002.3 and 1003.3 was done to coordinate the standard with the requirements of the ADA and ABA AG. The original intent of this text was to exempt faux balconies (balconies of very limited size that were not intended to be occupied) from the turning space or door approach requirements. The ADA Accessible unit requirements do not have a general exception for small balconies, but there is an exception in ADA and ABA AG Section 809.2.2 that eliminates the turning space requirement for small exterior spaces. The originally proposed change on this issue was modified and resulted in adding Exception 6 to Sections 1002.5 and 1003.5. Rather than addressing an arbitrary deck size of 30 inches, the exception applies to any space having a dimension less than the required maneuvering clearance that Section 404.2.3 would require on the exterior side of the door. The phrase "at other than the primary entrance door" was included to clarify that the exception was trying to address faux balconies and did not apply to an exterior location that could serve as the entry to the unit. This viewpoint is also supported by the inclusion of the word "balconies." By using the word "balconies," this section is intended to exempt small exterior balconies but would not apply to a deck that is associated with a ground-floor dwelling unit. Where a ground-floor unit is provided with an exterior deck, it does not face the same limitations that may restrict above-grade balconies from becoming larger.

1002.3.2, 1003.3.2

Turning Space

CHANGE TYPE: Modification

CHANGE SUMMARY: Two exceptions have been added to both the Accessible and Type A dwelling units to exempt certain spaces from turning space requirements: (1) bathrooms that do not have clearances at fixtures, and (2) where backing in or out would be acceptable as a minimum design requirement.

2009 STANDARD:

1002 Accessible Units

1002.3.2 Turning Space.   All rooms served by an accessible route shall provide a turning space complying with Section 304.

Exceptions:

1. A turning space shall not be required in toilet rooms and bathrooms that are not required to comply with Section 1002.11.2.

2. A turning space is not required within closets or pantries that are 48 inches (1220 mm) maximum in depth.

1003 Type A Units

1003.3.2 Turning Space.   All rooms served by an accessible route shall provide a turning space complying with Section 304.

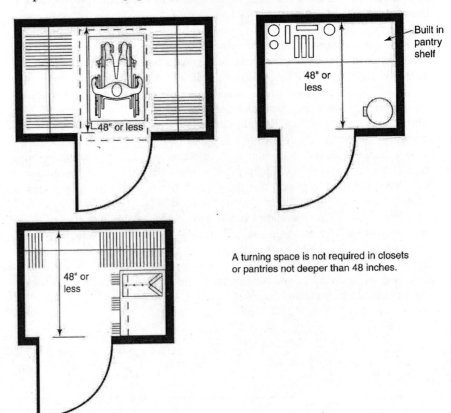

A turning space is not required in closets or pantries not deeper than 48 inches.

Exceptions:

1. A turning space is not required in toilet rooms and bathrooms that are not required to comply with Section 1003.11.2.

2. A turning space is not required within closets or pantries that are 48 inches (1220 mm) maximum in depth.

CHANGE SIGNIFICANCE: These added exceptions identify areas where a turning space is not required within Accessible and Type A units. There was no comparable change made to the Type B units because they do not require turning spaces within the rooms.

Exception 1 coordinates with the provisions of Sections 1002.11 and 1003.11 that require only one toilet and bathing room within the unit to be accessible. Any additional facilities would therefore be exempt from both the accessibility requirements and the need for a turning space within them. The requirement that only one bathroom within the unit is required to be accessible is consistent with the ADA and ABA AG requirements applicable to transient lodging guest rooms and residential dwelling units.

Exception 2 removes the requirement for the turning space when the room under consideration is a closet or a pantry that is not more than 48 inches in depth. The 48-inch depth was selected to coordinate with the clear floor space size and to limit the likelihood that persons would move far enough into the space where they could potentially become trapped if the door were to be closed behind them. Since closet doors seldom are equipped with a closer, it would be unlikely that the door would close or be closed when someone is able to proceed no more than 48 inches into the space. With such a shallow depth it is reasonable to permit a person to either back in or out of the space and not require a turning space to maneuver within the closet or pantry.

This exception allows for small pantries, storage closets, and deep clothes closets to have efficient use of the space as long as a person could get into the pantry/closet for access to shelves and rods. For example, a small kitchen pantry could have shelves on three sides as long as it had a 36-inch width between the side shelves and was not deeper than 48 inches to the face of the back shelves. Based on the discussions at the committee meeting, the 48-inch depth should be measured to the back wall of a closet and not to the edge of the hanging clothes that may be on the wall opposite the door. When dealing with a pantry, it would be reasonable to measure to the edge of the permanent shelves if they would be located low enough to prevent a 48-inch deep clear floor space within the pantry. Where the shelves are located at a height that would permit knee or toe clearances beneath them, it would be reasonable to measure to the wall and not to the edge of the shelf.

1002.5, 1003.5, 1004.5 continued

1004 Type B Units

1004.5 Doors and Doorways.   Doors and doorways shall comply with Section 1004.5.

1004.5.1 Primary Entrance Door.   The primary entrance door to the unit shall comply with Section 404.

Exception: ~~Maneuvering clearances required by Section 404.2.3 shall not be required on the unit side of the primary entrance door.~~ Storm and screen doors serving individual dwelling or sleeping units are not required to comply with Section 404.2.5.

CHANGE SIGNIFICANCE: The addition of most of the exceptions in the Accessible and Type A units will not result in any major changes but will instead simply clarify provisions that have generally been accepted or they will coordinate with other sections. For example, Exception 2 in Section 1002.5 removes the door-maneuvering clearance requirement from the toilet room or bathroom side of the door when that space is not required to be accessible. This coordinates with the requirement of Section 1002.11, which only requires one toilet or bathing facility in the unit to be accessible. This exception also matches a similar exception that had previously existed and applied to Type A dwelling units. Exception 6 replaces and modifies the provisions intended to address small balconies that were previously mentioned in Sections 1002.3 and 1003.3.

The exceptions that reference Section 404.2.5 eliminate the need to comply with the doors in series requirements. These exceptions address two issues related to the provisions of Section 404.2.5. First of all, if there is a small entrance hall or vestibule within a dwelling unit the turning space should not be required. This recognizes that within a dwelling unit there is less danger of entrapment within this area as there would be within an exterior vestibule in a public area or commercial building. Second, storm or screen doors as well as the communicating doors between guestrooms are not intended to be considered as doors in a series. The addition of the exceptions dealing with these closely spaced doors helps clarify this intent.

Although the exceptions dealing with storm/screen doors and the communicating doors between dwelling units only refer to Section 404.2.5, it would be illogical to apply the maneuvering clearances of Section 404.2.3 separately to each door and the space between them. When the doors are placed closely together as these types of doors are, it is appropriate to consider both of the doors as being a single doorway. If they are viewed as separate doors, then the code would generally require a 48-inch front approach between the two doors and a 12-inch latch side clearance could be required.

For information related to Exception 6 of Sections 1002.5 and 1003.5, see the discussion covering Sections 1002.3 and 1003.3 earlier in this book. The addition of Exception 6 was made as a part of the revisions to those two sections allowing for small balconies.

The most significant of all of these changes is the revision found in Section 1004.5.1 and applicable to the Type B dwelling units. Previously this exception eliminated the requirement for the maneuvering clearances on the unit side of the door. That exception was deleted and therefore the door-maneuvering clearances provision of Section 404.2.3 is now required both inside and outside of the unit on the primary entry door. Requiring the maneuvering space on the interior side of the primary entry door was done primarily to address the concern that this door is used as the means of egress from the unit and that if the space is not provided to properly approach the door, then quick access to the door during an emergency may be affected. While other doors within the unit are not required to provide maneuvering clearances, those doors can typically be removed by the occupants if needed. The option of removing the primary entry door typically does not exist due to security or weather issues.

The principal reasons that this change to the Type B units is such a significant issue is because the A117.1 will differ from the FHA requirements and because of the additional floor space that this requirement may impose on some units. HUD's Fair Housing Accessibility Guidelines (FHAG) does not require maneuvering space on the interior side of the primary entry door. Removing this exception from the standard will therefore make the requirements of the A117.1 standard more restrictive than the FHA. There could also be an additional cost impact depending on the door approach and the unit's configuration. For example, assuming a forward approach, besides the 18-inch latch side clearance, a depth of 60 inches must extend from the face of the door. This means a minimum 4-1/2-foot-by-5-foot foyer or other space would be required on the interior side of the door, whereas previously a 3-foot-wide hall could have run directly to the door.

1002.9, 1003.9, 1004.9

Operable Parts

CHANGE TYPE: Modification

CHANGE SUMMARY: Electrical panel boards have been added into the elements that are regulated under the operable parts section. Revisions to clarify some of the existing exceptions and the added exceptions for reset buttons and shut-offs will provide better direction and consistent interpretations when applying the requirements.

2009 STANDARD:

1002 Accessible Units

1002.9 Operable Parts. Lighting controls, electrical panelboards, electrical switches and receptacle outlets, environmental controls, appliance controls, operating hardware for operable windows, plumbing fixture controls, and user controls for security or intercom systems shall comply with Section 309.

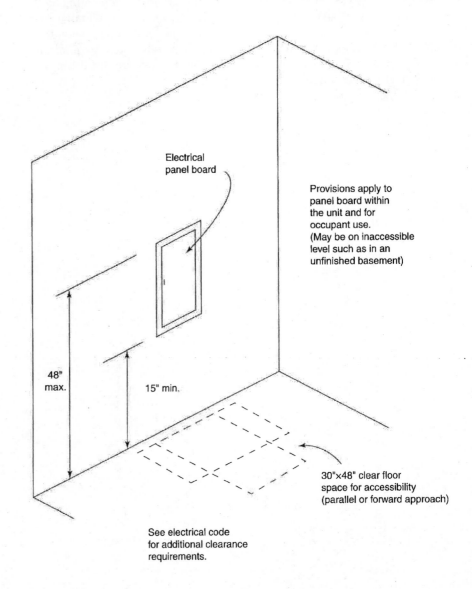

Electrical panel board

Provisions apply to panel board within the unit and for occupant use. (May be on inaccessible level such as in an unfinished basement)

48" max.

15" min.

30"x48" clear floor space for accessibility (parallel or forward approach)

See electrical code for additional clearance requirements.

Exceptions:

1. Receptacle outlets serving a dedicated use.

2. ~~One receptacle outlet shall not be required to comply with Section 309 where all of the following conditions are met:~~
   ~~(a) the receptacle outlet is above a length of countertop that is uninterrupted by a sink or appliance;~~
   ~~(b) at least one receptacle outlet complying with Section 1002.9 is provided for that length of countertop; and~~
   ~~(c) all other receptacle outlets provided for that length of countertop comply with Section 1002.9.~~

2. <u>Where two or more receptacle outlets are provided in a kitchen above a length of countertop that is uninterrupted by a sink or appliance, one receptacle outlet shall not be required to comply with 309.</u>

3. Floor receptacle outlets.

4. HVAC diffusers.

5. Controls mounted on ceiling fans.

6. Where redundant controls other than light switches are provided for a single element, one control in each space shall not be required to be accessible.

7. <u>Reset buttons and shut-offs serving appliances, piping, and plumbing fixtures.</u>

8. <u>Electrical panelboards shall not be required to comply with Section 309.4.</u>

1003 Type A Units

1003.9 Operable Parts.   Lighting controls, <u>electrical panelboards,</u> electrical switches and receptacle outlets, environmental controls, appliance controls, operating hardware for operable windows, plumbing fixture controls, and user controls for security or intercom systems shall comply with Section 309.

Exceptions:

1. Receptacle outlets serving a dedicated use.

2. ~~One receptacle outlet is not required to comply with Section 309 where all of the following conditions are met:~~
   ~~(a) the receptacle outlet is above a length of countertop that is uninterrupted by a sink or appliance; and~~
   ~~(b) at least one receptacle outlet complying with Section 1003.9 is provided for that length of countertop; and~~
   ~~(c) all other receptacle outlets provided for that length of countertop comply with Section 1003.9.~~

2. <u>Where two or more receptacle outlets are provided in a kitchen above a length of countertop that is uninterrupted by a sink or appliance, one receptacle outlet shall not be required to comply with Section 309.</u>

1002.9, 1003.9, 1004.9 continues

1002.9, 1003.9, 1004.9 continued

3. Floor receptacle outlets.

4. HVAC diffusers.

5. Controls mounted on ceiling fans.

6. Where redundant controls other than light switches are provided for a single element, one control in each space shall not be required to be accessible.

7. Reset buttons and shut-offs serving appliances, piping, and plumbing fixtures.

8. Electrical panelboards shall not be required to comply with Section 309.4.

1004 Type B Units

1004.9 Operable Parts. Lighting controls, electrical switches and receptacle outlets, environmental controls, electrical panelboards, and user controls for security or intercom systems shall comply with Sections 309.2 and 309.3.

Exceptions:

1. Receptacle outlets serving a dedicated use.

2. One receptacle outlet is not required to comply with Sections 309.2 and 309.3 where all of the following conditions are met:
   (a) the receptacle outlet is above a length of countertop that is uninterrupted by a sink or appliance; and
   (b) at least one receptacle outlet complying with Section 1004.9 is provided for that length of countertop; and
   (c) all other receptacle outlets provided for that length of countertop comply with Section 1004.9.

2. Where two or more receptacle outlets are provided in a kitchen above a length of countertop that is uninterrupted by a sink or appliance, one receptacle outlet shall not be required to comply with Section 309.

3. Floor receptacle outlets.

4. HVAC diffusers.

5. Controls mounted on ceiling fans.

6. Controls or switches mounted on appliances.

7. Plumbing fixture controls.

8. Reset buttons and shut-offs serving appliances, piping, and plumbing fixtures.

9. Where redundant controls other than light switches are provided for a single element, one control in each space shall not be required to be accessible.

10. Within kitchens and bathrooms, lighting controls, electrical switches and receptacle outlets are permitted to be located over cabinets with countertops 36 inches (915 mm) maximum in height and 25-1/2 inches (650 mm) maximum in depth.

CHANGE SIGNIFICANCE: A number of new exceptions and revisions to the existing provisions will affect what controls must comply with the operable parts requirements of Section 309 or exempt certain aspects such as the 5-pound operable force limit.

Electrical panelboards (circuit breaker boxes) have been added to the base paragraph in the sections for Accessible, Type A, and Type B units. It is important to note that the inclusion of panelboards into these sections does not mean that they are required to be placed within the unit but simply that they are accessible where they are installed within the units and are available for operation or use by the occupants. The location and additional access requirements for panelboards are found in the National Fire Protection Association's *National Electrical Code* (NFPA 70).

In the Accessible and Type A units, Exception 8 will exempt the panelboards as well as the circuit breakers or fuses within them from the operation requirements of Section 309.4. This means the panelboards need to have a clear floor space in front of them and be within the reach range height requirements per Sections 309.2 and 309.3. For the Type B units the base paragraph only requires compliance with Sections 309.2 and 309.3; therefore an exception similar to that of the Accessible and Type A units is not needed. Because panelboards and circuit breakers are not regulated by HUD's Fair Housing requirements, the A117.1 provisions will make the standard's Type B units more restrictive than the federal law in this aspect. However, because the ADA and ABA AG does regulate the electrical panel and circuit breakers, the changes for the Accessible and Type A units will help to harmonize the requirements for those units that must comply with both documents.

Exception 2 in this section has been replaced for each of the three types of dwelling units. The replacement language is taken from ADA and ABA AG Section 205.1, Exception 3. This new language will not change the provisions but provides better clarity in a concise format.

The new exception for reset buttons and shut-offs is intended to allow electrical switches and various shut-offs to be exempt from all of the requirements of Section 309. Inclusion of this new exception is a clarification that these controls were not typically considered as being regulated. Providing the exception provides clear guidance, these controls are not regulated by the standard. These types of switches or controls are not intended for everyday usage but are for protecting appliances or allowing them to be disconnected to be serviced. They may even be built in as a part of the appliance or fixture. This exception will eliminate the reach ranges, clear floor space, 5-pound force, and the no tight pinching or grasping requirements from applying. Some of the more common examples that may be found within a dwelling unit would include an electrical reset switch on the bottom of a garbage disposal unit or the water shut-off valves that may be found on water supply lines inside of a base cabinet or beneath a sink, lavatory, or water closet.

Exception 10 for the Type B units was added to address the fact that the FHA permits the use of standard 36-inch-high counters and also permits the standard 24-inch-depth cabinet to have a countertop equipped with a lip or edge extending beyond the cabinet. Since many of the FHA documents state that it was specifically the intent of that law to permit the use of standard-size cabinets, this exception is needed to coordinate the standard with the FHA. Because the reach range requirements of

1002.9, 1003.9, 1004.9 continues

118

1002.9, 1003.9, 1004.9 continued

A117.1 Section 308 are based on a maximum obstruction height of 34 inches and a maximum reach depth of 24 inches, the standard would be more restrictive than the FHA if this exception was not added. Although a similar exception permitting the 25-1/2-inch reach depth was not included for the Accessible and Type A units, the committee discussion recognized that the standard-depth cabinets and overhangs were routinely installed and accepted even though the countertop's overhang would technically move the reach depth beyond the 24-inch maximum that the standard specifies—the assumption being that the overhangs were being ignored or that the clear floor space is located adjacent to the base cabinet and therefore only 24 inches from the wall. The committee was not willing to modify the general 24-inch reach depth requirement without additional justification.

Exception 9 in Section 1004.9 provides an additional exemption that will permit redundant controls from needing to comply with the operable parts requirements. Where redundant controls are provided, the occupants would be assured of having access to operate the element since only one of the controls would be exempt.

CHANGE TYPE: Addition and Modification

CHANGE SUMMARY: Relocates all reinforcement requirements in front of the general bathroom requirements. Also relocates swing-up grab bar provisions from Chapter 6 to the Type B unit section. This is a companion change to those of Sections 604.5 and 607.4.

1002.11, 1003.11, 1004.11

Grab Bars and Shower Seat Reinforcement

2009 STANDARD:

1002 Accessible Units

1002.11 Toilet and Bathing Facilities. At least one toilet and bathing facility shall comply with Section 1002.11.2. All other toilet and bathing facilities shall comply with Section 1002.11.1.

1002.11.1 Grab Bars and Shower Seat Reinforcement. At fixtures in toilet and bathing facilities not required to comply with Section 1002.11.2, reinforcement in accordance with Section 1004.11.1 shall be provided.

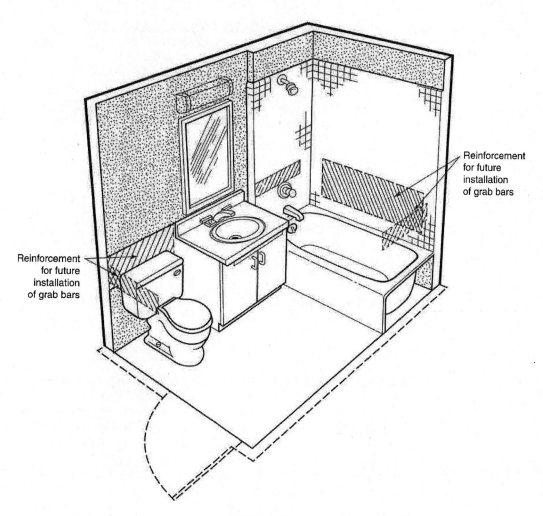

Reinforcement for future installation of grab bars

Reinforcement for future installation of grab bars

1002.11, 1003.11, 1004.11 continues

1002.11, 1003.11, 1004.11 continued

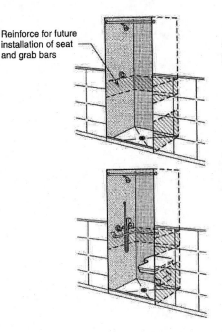

Reinforce for future installation of seat and grab bars

Exception: Reinforcement is not required where Type B units are not provided in the structure.

1003 Type A Units

1003.11 Toilet and Bathing Facilities. At least one toilet and bathing facility shall comply with Section 1003.11.2. All toilet and bathing facilities shall comply with Section 1003.11.1.

1003.11.1 ~~1003.11.4~~ **Grab Bar and Shower Seat Reinforcement.** Reinforcement shall be provided for the future installation of grab bars complying with Section 604.5 at water closets, grab bars complying with Section 607.4 at bathtubs, and for grab bars and shower seats complying with Sections 608.3, 608.2.1.3, 608.2.2.3, and 608.2.3.2 at shower compartments. ~~and shower seats at water closets, bathtubs, and shower compartments. Where walls are located to permit the installation of grab bars and seats complying with Sections 604.5, 607.4, 608.3 and 608.4, reinforcement shall be provided for the future installation of grab bars and seats meeting those requirements.~~

Exceptions:

1. At fixtures not required to comply with Section 1003. 11.2, reinforcement in accordance with Section 1004.11.1 shall be permitted.
2. Reinforcement is not required in a room containing only a lavatory and a water closet, provided the room does not contain the only lavatory or water closet on the accessible level of the dwelling unit.
3. Reinforcement for the water closet side-wall vertical grab bar component required by Section 604.5 is not required.
4. Where the lavatory overlaps the water closet clearance in accordance with the exception to Section 1003.11.2.4.4 reinforcement at the water closet rear wall for a 24 inch (610 mm) minimum length grab bar, centered on the water closet, shall be provided.

1004 Type B Units

1004.11 Toilet and Bathing Facilities. Toilet and bathing fixtures shall comply with Section 1004.11.

Exception: Fixtures on levels not required to be accessible.

~~**1004.11.2**~~ **1004.11.1 Grab Bar and Shower Seat Reinforcement.** Reinforcement shall be provided for the future installation of grab bars and shower seats at water closets, bathtubs, and shower compartments. Where walls are located to permit the installation of grab bars and seats complying with Sections ~~604.5~~ at water closets, grab bars complying with Section 607.4 at bathtubs, and for grab bars and shower seats complying with Sections 608.3, 608.2.1.3, 608.2.2.3, and

608.2.3.2 <u>at shower compartments</u>, reinforcement shall be provided for the future installation of grab bars and seats ~~meeting~~ <u>complying with</u> those requirements.

Exceptions:

1. ~~Reinforcement is not required~~ <u>In</u> a room containing only a lavatory and a water closet, <u>reinforcement is not required</u> provided the room does not contain the only lavatory or water closet on the accessible level of the unit.

2. <u>At water closets reinforcement for the side wall vertical grab bar component required by Section 604.5 is not required.</u>

3. <u>At water closets where wall space will not permit a grab bar complying with Section 604.5.2, reinforcement for a rear-wall grab bar 24 inches (610 mm) minimum in length, centered on the water closet shall be provided.</u>

4. <u>At water closets where a side wall is not available for a 42-inch (1065-mm) grab bar complying with Section 604.5.1, reinforcement for a side-wall grab bar, 24 inches (610 mm) minimum in length, located 12 inches (305 mm) maximum from the rear wall, shall be provided.</u>

5. <u>At water closets where a side wall is not available for a 42-inch (1065-mm) grab bar complying with Section 604.5.1 reinforcement for a swing-up grab bar complying with Section 1004.11.1.1 shall be permitted.</u>

6. <u>At water closets where a side wall is not available for a 42-inch (1065-mm) grab bar complying with Section 604.5.1 reinforcement for two swing-up grab bars complying with Section 1004.11.1.1 shall be permitted to be installed in lieu of reinforcement for rear wall and side wall grab bars.</u>

7. <u>In shower compartments larger than 36 inches (915 mm) in width and 36 inches (915 mm) in depth reinforcement for a shower seat is not required.</u>

1004.11.1 Swing-up Grab Bars. <u>A clearance of 18 inches (455 mm) minimum from the centerline of the water closet to any side wall or obstruction shall be provided where reinforcement for swing-up grab bars is provided. When the approach to the water closet is from the side, the 18 inches (455 mm) minimum shall be on the side opposite the direction of approach. Reinforcement shall accommodate a swing-up grab bar centered 15-3/4 inches (400 mm) from the centerline of the water closet and 28 inches (710 mm) minimum in length, measured from the wall to the end of the horizontal portion of the grab bar. Reinforcement shall accommodate a swing-up grab bar with a height in the down position of 33 inches (840 mm) minimum and 36 inches (915 mm) maximum. Reinforcement shall be adequate to resist forces in accordance with Section 609.8.</u>

Exception: <u>Where a water closet is positioned with a wall to the rear and to one side, the centerline of the water closet shall be 16 inches (405 mm) minimum and 18 inches (455 mm) maximum from the side wall.</u>

1002.11, 1003.11, 1004.11 continues

122

CHANGE SIGNIFICANCE: By relocating the reinforcement requirements in front of the general bathroom sections, it is clear that the blocking provisions are applicable to all water closets and bathing facilities within the individual units, not just the bathrooms with clearances at fixtures. Providing the blocking provisions at this one location allows them to be applicable to both water closets and to bathing facilities. This eliminates the need for the provisions to be duplicated within both of those sections. Because the FHA requires reinforcement for the future installation of grab bars in all bathrooms, providing this information in the Accessible and Type A units helps to ensure these units will be accepted as a more accessible alternative for the Type B units.

Many of these provisions are not new but are being relocated from Chapter 6 to Chapter 10. Moving these items to Chapter 10 allows the requirements for the bathrooms in dwelling units to be found in one location instead of needing to refer to Chapter 6 and perhaps overlooking one of the grab bar requirements.

While Accessible units do require the installation of all grab bars at the time of construction, they are only required in one toilet or bathing facility within the unit (Section 1002.11.2). Sections 1002.11 and 1002.11.1 therefore add the requirement for reinforcement for the non-accessible bathrooms within Accessible units where the unit has two or more bathrooms. This helps to keep the requirements for Accessible units consistent with those of the FHA and the Type B unit provisions of Section 1004.11. In buildings where Type B units would not be required, the exception will eliminate the need for the Accessible units to include reinforcement and follow the Type B provisions. An example of this is a two-bathroom suite in a hotel.

As mentioned in the change summary, this is a companion change to those of Sections 604.5 and 607.4, where many of these provisions were relocated from. As the discussion on the pages for those sections indicated, the grab bar requirements for the various dwelling unit types have been moved from Chapter 6 to the appropriate location in Chapter 10. This consolidates the requirements for the various units into the proper section and should also help clarify the application of the requirements. See the discussion covering Sections 604.5 and 607.4 related to the swing-up grab bars for an example of how this will clarify the requirements of the standard.

The swing-up grab bar requirements have been modified to better address the location of the bars in relationship to the water closet and also the direction of approach. Where a wall is provided to support the grab bar, a 16- to 18-inch clearance is appropriate. Where a swing-up grab bar is to be utilized, a clearance of 18 inches minimum is required. The additional language in Section 1004.11.1.1 will allow for this and indicate that the swing-up bar location should be on the far side from the approach.

Exception 7 in Section 1004.11.1 has been duplicated from the "Option A" shower compartment provisions of Section 1004.11.3.1.3.3 and not moved from Chapter 6.

CHANGE TYPE: Modification

CHANGE SUMMARY: Accessible units are permitted to have only one bathroom in a unit be accessible. The revisions will also allow either a bathtub or a shower to serve as the accessible bathing facility.

2009 STANDARD:

1002 Accessible Units

1002.11 Toilet and Bathing Facilities.   At least one toilet and bathing facility shall comply with Section 1002.11.2. All other toilet and bathing facilities shall comply with Section 1002.11.1.

1002.11.2 Accessible Toilet and Bathing ~~Facilities~~ Facilty.   At least one toilet and bathing ~~facilities~~ facility shall comply with Sections 603 ~~through 610~~. At least one lavatory, one water closet, and either a bathtub or shower within the unit shall comply with Sections 604 through 610. The accessible toilet and bathing fixtures shall be in a single toilet/bathing area, such that travel between fixtures does not require travel through other parts of the unit.

1002.11.2.2 Mirrors.   Mirrors above accessible lavatories shall have the bottom edge of the reflecting surface 40 inches (1015 mm) maximum above the floor.

1003 Type A Units

1003.11 Toilet and Bathing Facilities.   At least one toilet and bathing facility shall comply with Section 1003.11.2. All toilet and bathing facilities shall comply with Section 1003.11.1.

1002.11.2, 1003.11.2 continues

124

1002.11.2, 1003.11.2 continued

1003.11.2 ~~Toilet and Bathing Facilities. 1003.12.1~~ **General.** ~~All toilet and bathing areas shall comply with Section 1003.11.4.~~ At least one toilet and bathing facility shall comply with Section ~~1003.11~~ 1003.11.2. At least one lavatory, one water closet and either a bathtub or shower within the unit shall comply with Section ~~1003.11~~ 1003.11.2. The accessible toilet and bathing fixtures shall be in a single toilet/bathing area, such that travel between fixtures does not require travel through other parts of the unit.

1003.11.2.3 Mirrors. Mirrors above <u>accessible</u> lavatories shall have the bottom edge of the reflecting surface 40 inches (1015 mm) maximum above the floor.

CHANGE SIGNIFICANCE: These sections contain a number of changes that were made to help coordinate the requirements of the various types of dwelling units. Perhaps the most significant of the changes is that of Section 1002.11.2, where the requirements are now limited to applying to a single bathroom and single type of fixture within the unit. Previously if an Accessible dwelling unit had multiple toilet or bathing facilities within the unit, they all would have been regulated unless the jurisdiction's separate scoping document provided additional guidance. The added language also ensures that the accessible fixtures are located in the same area instead of allowing the possibility that the accessible water closet could be in one bathroom while the accessible lavatory or shower were located in a completely different bathroom of the unit. The language does not intend to limit the possibility that a water closet or shower could be located within a separate room for privacy as long as it is all a part of the same bathroom area. The revisions of Section 1002.11.2 were made to coordinate the Accessible units with the ADA and ABA AG.

One other important aspect of this change is the fact that Section 1002.11.2 will allow "either" a bathtub or a shower to be the accessible bathing element. This coordinates with the idea of only one bathing element being required to be accessible in all of the various types of units. Whether the accessible bathing fixture is a tub or a shower is a design choice. This choice will still exist even within a bathroom that has both a tub and a shower, a common amenity in many master suites today. So while the designer may elect to make both bathing fixtures accessible, the standard will only require compliance for one of them. This will allow for the freedom to design bathrooms with platform, sunken, or stand-alone tubs, or showers with two or more glass walls, as long as the other fixture in the bathroom meets accessibility provisions. When two lavatories are provided within a shared bathroom, only one lavatory will be required to be designed for accessibility. This is consistent with the application and intent of the ADA and ABA AG and also the Type A unit provisions of Sections 1003.11.2 and 1003.11.2.5. See the coverage of Section 1003.11.2.5 later in this book for additional discussion related to this topic.

See the page dealing with Section 603.3 for information related to the changes for mirrors.

The change to the Type A units in Section 1003.11.2 is simply deleting the text that requires blocking for the future installation of grab bars and shower seats. Since the reinforcement requirements have been relocated into a separate Section 1003.11.1 and the reinforcement is required for all of the toilet and bathing facilities, this redundant text in Section 1003.11.2 is no longer needed.

CHANGE TYPE: Modification

CHANGE SUMMARY: Clarifies that some of the "kitchen" requirements will also apply to kitchenettes that may occur within the units. The change will affect suites in facilities such as assisted living, dormitories, efficiency apartments, and in hotels that include a kitchenette.

2009 STANDARD:

1002 Accessible Units

1002.12 Kitchens and Kitchenettes.   Kitchens <u>and kitchenettes</u> shall comply with Section 804. At least one work surface, 30 inches (760 mm) minimum in length, shall comply with Section 902.

Exception: Spaces that do not provide a cooktop or conventional range shall not be required to provide an accessible work surface.

1003 Type A Units

1003.12 Kitchens and Kitchenettes.   Kitchens <u>and kitchenettes</u> shall comply with Section 1003.12.

1004 Type B Units

1004.12 Kitchens <u>and Kitchenettes</u>.   Kitchens <u>and kitchenettes</u> shall comply with Section 1004.12.

CHANGE SIGNIFICANCE: The inclusion of the wording "and kitchenettes" will help to clarify what requirements are applicable when an accessible unit in facilities such as assisted living, dormitories, efficiency apartments, or a hotel are provided with a kitchenette. Without the added language it previously resulted in a variety of opinions when a unit was provided with something such as a bar sink, refrigerator, or microwave but not a cooktop or oven. Because kitchenettes have become more common in dormitories, hotels, and other similar units where a full kitchen is not typically provided, this added language will provide the users with better direction and eliminate the debate regarding whether such spaces are regulated by the "kitchen" provisions.

Although the requirement does apply to all three types of units (Accessible, Type A, and Type B), the application of the new text will mostly affect sleeping units and will have less of an impact on a dwelling unit. This is because it is more common to include a kitchenette within a sleeping unit than it is to see a kitchenette included in an accessible dwelling unit that already contains a kitchen.

Because kitchenettes do not include a cooktop, conventional range, or oven, the requirements for kitchenettes will result in several reductions or exceptions from the general kitchen requirements. Some of these reductions can be found within the individual sections, such as the exception for the work surface in Section 1002.12, while others would be found elsewhere, such as Exception 1 to Section 606.2.

1002.12, 1003.12, 1004.12

Kitchens and Kitchenettes

1002.12, 1003.12, 1004.12 continues

126

1002.12, 1003.12, 1004.12 continued

A parallel approach to the sink in a kitchenette would be permitted versus the forward approach that would typically apply in the Accessible and Type A units. This is based on both Exceptions 1 and 6 in Section 606.2. This allowance for the parallel approach to the sink is clearly permitted for the Accessible units because of the reference from Section 1002.12 up to Section 804 and then from Section 804.4 to Section 606 and its subsection 606.2.

When dealing with the Type A units, however, the requirements are not as clear and, depending upon the interpretation, they may even result in those units being more restrictive than the Accessible units for certain requirements. As stated earlier, when dealing with an Accessible unit, Section 1002.12 clearly provides the exception for work spaces and also the reference to Section 804 that will result in the parallel approach to the sink. The Type A unit requirements of Section 1003.12 do not provide an equivalent exception for the work surfaces or for allowing the sink to have a parallel approach. Specifically, the problem can be seen in the work surface requirements of the Type A units where Sections 1003 and 1003.12.3 do not contain an exception similar to that of Section 1002.12. A similar problem also exists with the clearance requirements of Section 1003.12.1 and 1004.12.1, which do not contain an exception similar to that found in Section 804.2. Remember that Type B units do not have a requirement for a kitchen work surface and already allow for a side approach to the sink, so there should not be the same issue for Type B units.

Because of these inconsistencies, users must decide to either be code literal and make the Type A units more restrictive and provide better access than required for the Accessible units, or they must use their judgment to permit the Type A units to use the exceptions that are allowed for an Accessible unit. Because an Accessible unit is considered as the higher level of accessibility, I believe that granting these needed reductions for kitchenettes in Type A units is a reasonable solution even if it cannot be supported by the specific language of the standard. Scoping provisions for Type A units in the IBC are limited to large apartment buildings, convents, and monasteries, so this situation should be limited to a relatively small number of locations. However, if this situation does arise, this would be an important item to discuss with the enforcing agency prior to proceeding too far in the design or construction process.

CHANGE TYPE: Modification

CHANGE SUMMARY: Revises scoping to tie the window requirements to required windows versus where operable windows are provided. The sections are modified so the operating force requirements do not apply to the window's operation.

1002.13, 1003.13
Windows

2009 STANDARD:

1002 Accessible Units

1002.13 Windows. ~~Where operable windows are provided, at least one window in each sleeping, living, or dining space shall have operable parts complying with Section 1002.9. Each required operable window shall have operable parts complying with Section 1002.9.~~ <u>Windows shall comply with Section 1002.13.</u>

<u>1002.13.1 Natural Ventilation.</u>   <u>Operable windows required to provide natural ventilation shall comply with Sections 309.2 and 309.3.</u>

1002.13.2 Emergency Escape.   <u>Operable windows required to provide an emergency escape and rescue opening shall comply with Section 309.2.</u>

1003 Type A Units

1003.13 Windows. ~~Where operable windows are provided, at least one window in each sleeping, living, or dining space shall have operable parts complying with Section 1003.9. Each required operable window shall have operable parts complying with Section 1003.9.~~ <u>Windows shall comply with Section 1003.13.</u>

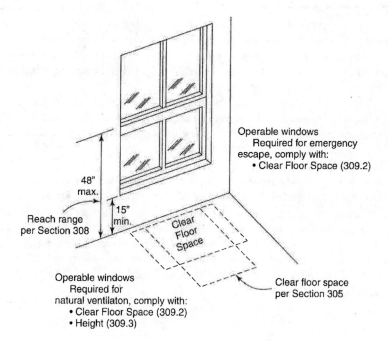

1002.13, 1003.13 continues

1002.13, 1003.13 continued

1003.13.1 Natural Ventilation. <u>Operable windows required to provide natural ventilation shall comply with Sections 309.2 and 309.3.</u>

1003.13.2 Emergency Escape. <u>Operable windows required to provide an emergency escape and rescue opening shall comply with Section 309.2.</u>

CHANGE SIGNIFICANCE: Perhaps the most significant changes in these sections are found within the text that has been deleted. First of all, the scoping of the window requirements previously applied to locations "where operable windows are provided" and then required "at least one window" in each of several spaces to comply. The new text changes the scoping so that it applies to operable windows that are *required* to provide natural ventilation or emergency escape and rescue. The requirements will therefore apply in fewer locations and it would be permissible to have operable windows that do not meet these provisions provided they are not required for one of the two stated reasons.

By the same language, it is possible that there may be more windows regulated when the windows are for ventilation purposes than what would have previously been required. Under the previous language, "at least one" window was required in specific locations. With the new language, any window that is required to meet the natural ventilation requirements would be regulated. For example, if a room required two windows to provide the proper amount of ventilation, then both of the windows would need to comply. Previously, only one of the windows would have been regulated. In addition, if a window in a bathroom was provided for required natural ventilation, then the window would clearly be regulated under the new language while it was debatable whether this was a "living" space under the previous edition.

The other substantive change to the sections is the fact that previously the referenced sections would have imposed the clear floor space, height, and operation requirements from Section 309. The new language will only refer to the clear floor space provisions of Section 309.2 and for the natural ventilation windows will also impose the reach range limitations of Section 309.3 and Sections 308. The separate requirements for the two window types are clearly shown in the two subsections for both the Accessible and Type A units. Type B units do not require windows to meet any accessibility requirement.

The window requirements of the A117.1 standard are more restrictive and do not coordinate with the ADA and ABA AG. The A117.1 requirements are not overly restrictive since they are primarily going to require an accessible route and a clear floor space to the required windows. The ADA and ABA AG exempts windows in residential dwelling units due to a concern for child safety at the windows. The assumption in the federal law is that the residents may either request a reasonable accommodation from the owner to provide accessible window actuators or they may modify the unit at their own expense. This difference with the federal law and also because of the fact that the IBC does impose some other requirements on the emergency escape and rescue openings help show one of the reasons that the requirements have been reduced from the 2003 edition of the standard.

CHANGE TYPE: Modification

CHANGE SUMMARY: Primarily this is a companion change to the revision in Section 804.5 eliminating the kitchen storage requirements.

2009 STANDARD:

1002 Accessible Units

1002.14 Storage Facilities. Where storage facilities are provided, ~~they~~ at least one of each type shall comply with Section 905.

Exception: Kitchen cabinets shall not be required to comply with Section ~~804.5~~ 1002.14.

1003 Type A Units

~~**1003.12.5 Kitchen Storage.** A clear floor space, positioned for a parallel or forward approach to the kitchen cabinets, shall be provided.~~

1003.14 Storage Facilities. Where storage facilities are provided, ~~they~~ at least one of each type shall comply with Section 905 ~~1003.14~~.

Exception: Kitchen cabinets shall not be required to comply with Section ~~1003.12.5~~ 1003.14.

~~**1003.14.1 Clear Floor Space.** A clear floor space complying with Section 305, positioned for a parallel or forward approach, shall be provided at each storage facility.~~

1002.14, 1003.12.5, 1003.14

Storage Facilities

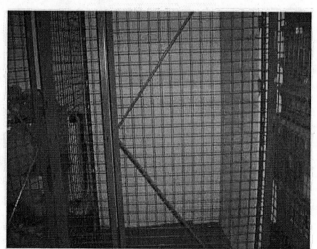

Courtesy United Spinal Association

1002.14, 1003.12.5, 1003.14 continues

1002.14, 1003.12.5,
1003.14 continued

~~**1003.14.2 Height.** A portion of the storage area of each storage facility
shall comply with at least one of the reach ranges specified in Section 308.~~

~~**1003.14.3 Operable Parts.** Operable parts on storage facilities shall
comply with Section 309.~~

CHANGE SIGNIFICANCE: The revisions eliminate the requirement that
kitchen cabinets within Accessible and Type A units comply with accessible storage provisions. Type B units do not have any accessible storage
requirements. This portion of the change is in recognition that kitchen
cabinets are extremely difficult to make fully accessible. The lower shelf
in the base cabinets is below the reach range while the upper cabinets
are typically beyond the reach range. Standard appliances in the lower
cabinets (i.e., range, dishwasher, garbage disposal) eliminate lower cabinet options, so there really isn't a percentage number that can be considered reasonable for all sizes of kitchens. However, just by being able to
move within the kitchen (via the appliance clearance requirements and
width between counters) would provide access to most base cabinets,
drawers, and counter storage. Adding user-friendly items such as pull-out
shelves in lower cabinets or a lazy Susan in corner cabinets is fairly easy
to accomplish. Using extended reaching tools will allow limited access to
upper cabinets. While the person with limited reach range might not be
able to get to all the cabinets, his or her family can still utilize the additional cabinets permitted. Therefore it is logical for the kitchen storage
requirements to have been removed. For related discussion, see the page
of this book covering Section 804 and the portion dealing with the deletion of Section 804.5.

Storage facilities, other than those in kitchen cabinets, have also been
modified so that "at least one of each type" is made accessible. The types
of storage locations that may be affected by this requirement are tenant
storage lockers/closets that may be located in a common area of an apartment building, an outdoor storage closet on the balcony of the dwelling
unit, or a private storage closet within a garage.

CHANGE TYPE: Addition

CHANGE SUMMARY: Requirements for beds in Accessible units have been added to the standard. Provisions address the scoping, clear floor space around the bed, and require an open bed frame to permit the use of a lift.

2009 STANDARD:

1002 Accessible Units

1002.15 Beds.   In at least one sleeping area, a minimum of five percent, but not less than one bed shall comply with Section 1002.15.

1002.15.1 Clear Floor Space.   A clear floor space complying with Section 305 shall be provided on both sides of the bed. The clear floor space shall be positioned for parallel approach to the side of the bed.

Exception: Where a single clear floor space complying with Section 305 positioned for parallel approach is provided between two beds, a clear floor space shall not be required on both sides of the bed.

1002.15.2 Bed Frames.   At least one bed shall be provided with an open bed frame.

1002.15

Beds in Accessible Units

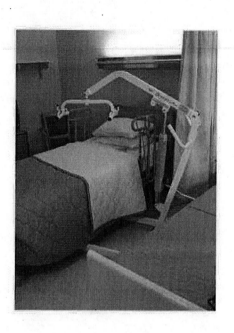

CHANGE SIGNIFICANCE: The intent of all three of these sections is to make at least one of the beds within every Accessible unit accessible to the occupants. The portions of this text that will be considered as the most substantial changes will be the scoping of Section 1002.15 and the open bed frame requirements of Section 1002.15.2. The clear floor space requirements of Section 1002.15.1 harmonize with the transient lodging guestroom requirements of the ADA and ABA AG and are similar to requirements that exist in Appendix E of the IBC.

In the scoping of Section 1002.15 the phrase "in at least one sleeping area" is important to permit the requirement to apply to a number of different situations where an Accessible unit may be provided. The primary intent of that phrase is similar to the requirement for toilet and bathing facilities (Section 1002.11.2), where "at least one" facility within the Accessible unit is regulated. In a unit where there are multiple bedrooms or sleeping areas, only one of them would require the complying bed. On the other hand, the language also does not allow the provision to be applied across the entire building but is instead applied to each unit. For example, in a typical college dormitory or hotel, the bed requirements would not apply to only a single sleeping area or unit in the structure but would be applied individually within each Accessible unit within the building. The IBC provides the scoping for the number of Accessible units while A117.1 Section 1002.15 addresses the number of sleeping areas *within* the unit that are regulated.

In a large dormitory or barracks-type setting where all of the beds are in the same room or area the requirement for "a minimum of five percent" means that if there were more than 20 beds within the space, more than one of them would be regulated.

1002.15 continues

132

1002.15 continued The requirement of Section 1002.15.2 for an open bed frame is intended to allow the use of a bed lift such as a Hoyer lift to be used to transfer either to or from the bed. Having the open bed frame allows the legs of the lift to extend under the bed and provides a greater base of support for the lift so that it is stable and secure during the transfer. Although the language of Section 1002.15.2 could be viewed as meaning that any one bed within the unit could have the open frame, the language of Section 1002.15 should be used so that the bed with the open frame is also the bed served by the clear floor space. This is a reasonable interpretation since it will allow the person to maneuver adjacent to the bed before using the lift or allow an assistant to maneuver the lift into the clear floor space before transferring the person back to the bed.

1003.11.2.4.4
Clearance Overlap for Toilet and Bathing Facilities

CHANGE TYPE: Modification

CHANGE SUMMARY: The exception is revised so the depth of the lavatory adjacent to the water closet is limited. Previously the depth of the obstruction was not addressed.

2009 STANDARD:

1003 Type A Units

1003.11.2.4.4 <u>Clearance</u> Overlap.   The required clearance around the water closet shall be permitted to overlap the water closet, associated grab bars, paper dispensers, coat hooks, shelves, accessible routes, clear floor space required at other fixtures, and the wheelchair turning space. No other fixtures or obstructions shall be located within the required water closet clearance.

Exception: A lavatory <u>measuring 24 inches (610 mm) maximum in depth and</u> complying with Section 1003.11.2.2 shall be permitted on the rear wall 18 inches (455 mm) minimum from the centerline of the water closet <u>to the side edge of the lavatory</u> where the clearance at the water closet is 66 inches (1675 mm) minimum measured perpendicular from the rear wall.

CHANGE SIGNIFICANCE: The addition of the 24-inch maximum depth for the lavatory provides a limitation that did not previously exist for the lavatory in the Type A unit. Although the standard did allow a lavatory adjacent to the water closet, there was no information that provided guidance on the maximum depth of the obstruction. Without a limit the lavatory or the cabinet that it was installed in could have possibly been deep enough that it would have restricted or blocked access to the water closet. Providing a specific limit on the obstruction will eliminate potential debate and make design and enforcement easier. While the exception specifically says "lavatory," most lavatories are installed in a counter. In a Type A unit, cabinets can be installed below as long as they are removable.

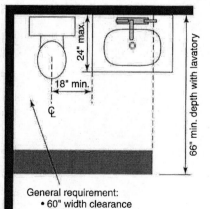

Clearance overlap exception:
   Lavatory permitted in
   clearance provided
 • Lavatory depth 24" maximum
 • 18" minimum from center line
   of water closet
 • Clearance depth increased to
   66" minimum

General requirement:
 • 60" width clearance
 • 56" depth clearance
 • No other fixtures or obstructions in clearance

1003.11.2.4.4 continues

1003.11.2.4.4 continued

The 24-inch maximum depth was selected to match the depth that is allowed in Section 1004.11.3.1.2.2.4 for Type B units where a vanity or other obstruction overlaps the clear floor space for the water closet. Providing a consistent limit on the depth of the lavatory or vanity will help keep the Type A and Type B units consistent so that a Type A unit will be allowed as an alternate for a Type B unit. A question still remaining is if 24 inches is the depth of the counter or the depth of the cabinet below. The FHA specifically allows for 25-1/2-inch-deep countertops on 24-inch-deep cabinets. Once the cabinets are removed, with the knee and toe clearance available, are they still an obstruction to providing access to the water closet?

The other change within the exception is simply a clarification of where the measurement is to be made for clearance between the edge of the lavatory/vanity and the water closet. The change in the section title coordinates with changes made to Sections 1003.11.2.4.2 and 1003.11.2.4.3.

CHANGE TYPE: Addition

CHANGE SUMMARY: This new section provides scoping for the bathtub and shower sections and clarifies that either fixture can be the accessible bathing fixture when both are provided in the same bathroom.

2009 STANDARD:

1003 Type A Units

1003.11.2.5 Bathing Fixtures. The accessible bathing fixture shall be a bathtub complying with Section 1003.11.2.5.1 or a shower compartment complying with Section 1003.11.2.5.2.

1003.11.2.5
Bathing Fixtures

1003.11.2.5 continues

136

1003.11.2.5 continued   **1004 Type B Units**

Option A bathroom

1004.11.3.1.3 Bathing <u>Fixtures</u> ~~Facilities~~. ~~Where a bathtub or shower compartment is provided it shall conform with Section 1004.11.3.1.3.1, 1004.11.3.1.3.2, or 1004.11.3.1.3.3.~~ <u>Where provided, a bathtub shall comply with Sections 1004.11.3.1.3.1 or 1004.11.3.1.3.2 and a shower compartment shall comply with Section 1004.11.3.1.3.3.</u>

Option B bathroom

1004.11.3.2.3 Bathing <u>Fixtures</u> ~~Facilities~~. ~~Where either a bathtub or shower compartment is provided, it shall conform with Section 1004.11.3.2.3.1 or 1004.11.3.2.3.2.~~ <u>The accessible bathing fixture shall be a bathtub complying with Section 1004.11.3.2.3.1 or a shower compartment complying with Section 1004.11.3.2.3.2.</u>

CHANGE SIGNIFICANCE: This new Section 1003.11.2.5 serves two main purposes. The first purpose of this section is to provide the scoping for the bathtub and shower requirements that follow this section in the standard. The second and more important is that this coordinates with the idea of only one bathing element being required to be accessible in all of the various types of units.

The idea that only one bathing fixture is regulated within a bathroom that has both a tub and a shower is not new. This concept is included in the general toilet and bathing requirements of Section 1003.11.2, where it indicates that *either* a bathtub or a shower within the unit shall comply. Similar language has also been placed into Section 1002.11.2 for the Accessible units. See the related discussion covering Section 1002.11.2 earlier in this book.

When dealing with the Type B units, this concept of allowing either a tub or a shower is not as clear and depends on both HUD's interpretation of the Fair Housing requirements, including the Fair Housing Accessibility Guidelines (FHAG), and on whether an "Option A" or "Option B" bathroom is used. In Section 1004.11.3.1.3 the Option A bathing fixture requirements use the word "and" as the connector between the bathtub and shower requirements. This section also starts out by indicating that "where provided," the bathing fixtures must comply with the specific sections. Therefore where both a bathtub and a shower are provided, the interpretation would be that both of them must be accessible. This interpretation, although seemingly clear in the standard and consistent with HUD's opinion, will make the Type B, Option A bathroom units with both a tub and a shower in a single bathroom more restrictive than the ADA and ABA AG and also the Accessible and Type A units from the A117.1 standard. If the Option B bathrooms are used within the Type B unit, then Section 1004.11.3.2.3 would be consistent with the Accessible and Type A units and require only a single bathing fixture to comply. This is based on the fact that the text states "*The* accessible bathing fixture shall be a bathtub . . . *or* a shower. . . ."

If designers do not plan on making both fixtures accessible in a Type B unit, Option A bathroom, I would suggest they check with HUD when placing both a tub and a shower in the same bathroom. As stated earlier, the literal interpretation of the standard and HUD's interpretations are very clear. However, in my opinion, if only one bathing fixture is regulated in the Accessible units, Type A units and Type B units with an Option B bathroom, as well as units covered by the ADA and ABA AG, it is inconsistent that the Type B units with Option A bathrooms and their typically lower level of accessibility would or should require both the tub and shower to comply simply because they are both installed in the same bathroom. Since a designer could solve this issue by removing either the tub or the shower, is this not eliminating a design option? In addition, the FHA preamble states that it is the intent for Option B bathrooms to be more accessible than Option A bathrooms. Therefore the interpretation that both bathing fixtures must be accessible in the Option A bathroom is inconsistent with that stated intent. The reasonableness of the HUD interpretation is stretched even farther if there are multiple bathrooms in the unit and the requirements of Section 1004.11.3 are considered. In that situation a single bathroom built to the Option B bathroom provisions would require only one bathing fixture in one bathroom to be accessible. On the other hand, the Option A bathroom provisions would mean that both the bathtub and the shower in every bathroom within the unit would be regulated. This again is inconsistent with the stated intent that Option B bathrooms are more accessible.

One design option that users should be aware of is the fact that if they make just a few small changes in their Option A bathroom they can get away from this requirement for both the shower and tub to be accessible. Based on the limited differences between the Option A and Option B bathroom requirements, it would appear that by using a 34-inch-high lavatory (Section 1004.11.3.2.1.1) and then providing access to the shower *or* the tub, you have then created an Option B bathroom that would require only one of the bathing fixtures to be accessible (Section 1004.11.3.2.3) and would also eliminate the access requirements for all other bathrooms within the unit (Section 1004.11.3). This simple design change may make compliance much easier. While I don't see how a lower-height lavatory affects the access to the tub and shower, it does seem to be an easy trade-off to make.

1003.12.4.1

Clear Floor Space for Kitchen Sinks

Kitchen in Type A unit

Clear floor space is not required to be centered on the sink bowl

CHANGE TYPE: Modification

CHANGE SUMMARY: Deletes the requirement for the clear floor space to be centered on the kitchen sink. A consistent format is provided for the future removal of base cabinets to allow a more accessible option of a forward approach to sinks or lavatories.

2009 STANDARD:

1003 Type A Units

1003.12.4.1 Clear Floor Space. A clear floor space, positioned for a forward approach to the sink, shall be provided. Knee and toe clearance complying with Section 306 shall be provided. ~~The clear floor space shall be centered on the sink bowl.~~

Exceptions:

1. The requirement for knee and toe clearance shall not apply to more than one bowl of a multi-bowl sink.
2. Cabinetry shall be permitted to be added under the sink, provided the following criteria are met:
   (a) The cabinetry can be removed without removal or replacement of the sink,
   (b) The floor finish extends under ~~such~~ the cabinetry, and
   (c) The walls behind and surrounding the cabinetry are finished.

CHANGE SIGNIFICANCE: The requirement for the clear floor space to be centered on the bowl of the sink has been deleted. Centering of the space is important for a parallel approach but not for a forward approach. While it generally may be the best practice to center the clear floor space, it may also create difficulties for certain types of sinks or unnecessarily limit the design options. The previous requirement for the clear floor space to be centered on the kitchen sink in a Type A unit would appear to have been unjustified or unnecessary given that the centering requirement did not apply to lavatories in the bathroom or to the kitchen sink in an Accessible unit or in other kitchens (Section 606.2).

Given that many kitchen sinks may include a garbage disposal unit beneath the sink, drains and faucets at varying locations, or that some sinks come with multiple bowls and even different sizes of bowls, it would appear that the requirement for centering may not always be possible with a forward approach or that it may not provide the best access. Eliminating this centering requirement will make the Type A units consistent with the Accessible units and the general kitchen requirements of Section 804 that have apparently provided adequate access in the past without having this requirement.

Although it is not a change in the technical requirements, the format used in Exception 2 is important because it is now used throughout the standard. For those locations where a forward-approach lavatory is permitted to have removable cabinets installed beneath them, the standard will use this format for consistency. Some of the examples of other sections using this style for the exception are Sections 1003.11.2.5.1, 1003.11.2.5.2, and 1004.11.3.1.1.

CHANGE TYPE: Modification

CHANGE SUMMARY: A new exception clearly addresses that only a single lavatory needs to be accessible within the same bathroom in a Type B unit, Option A bathroom.

2009 STANDARD:

1004 Type B Units

1004.11.3.1 Option A.   Each fixture provided shall comply with Section 1004.11.3.1.

Exceptions:

1. Where multiple lavatories are provided in a single toilet and bathing area such that travel between fixtures does not require travel through other parts of the unit, not more than one lavatory is required to comply with Section 1004.11.3.1.

2. A lavatory and a water closet in a room containing only a lavatory and water closet, provided the room does not contain the only lavatory or water closet on the accessible level of the unit.

CHANGE SIGNIFICANCE: The addition of this exception helps to resolve a question that previously existed regarding whether a bathroom with multiple lavatories required all the fixtures to be accessible or just one. The addition of this exception continues the committee's other changes that have helped to clarify that only one of each type of fixture is required to be accessible. This issue has been discussed in other sections in this book and was clearly accepted previously for the Type A units (see Section 1003.11.2).

This concept of applying the accessible requirements to "one of each type of fixture" is clearly spelled out for the Type B units, Option B bathrooms in Section 1004.11.3.2. However, because the language for the Type B

1004.11.3.1

Multiple Lavatories in Type B Unit, Option A Bathrooms

Type B Unit - Option A Bathroom

General requirement: Each fixture must comply with Option A requirements

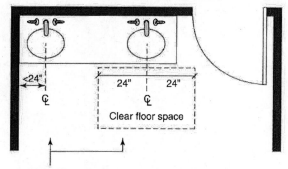

Exception:
Where multiple lavatories are provided, not more than one is
required to provide a clear floor space centered on the lavatory

1004.11.3.1 continues

140

units, Option A bathroom requirements stated that "*each* fixture provided" had to comply, the provisions were typically imposed on both lavatories where multiple lavatories were provided.

The elimination of this requirement for multiple lavatories will help where one of the lavatories is pushed closer than 24 inches to a wall and would not allow for the clear floor space to be centered on the fixture. Alternatively, the designer could be forced to eliminate one of the lavatories, thus eliminating a design option for clients—not the best idea in master bathrooms or bathrooms shared by large families.

CHANGE TYPE: Modification

CHANGE SUMMARY: This change clarified and coordinated the maneuvering clearances around the water closets. Provisions are in smaller, easier-understand sections that are addressing a single issue.

2009 STANDARD:

1004 Type B Units

~~**1004.11.3.1.2 Water Closet.** The lateral distance from the centerline of the water closet to a bathtub or lavatory shall be 18 inches (455 mm) minimum on the side opposite the direction of approach and 15 inches (380 mm) minimum on the other side. The lateral distance from the centerline of the water closet to an adjacent wall shall be 18 inches (455 mm). The lateral distance from the centerline of the water closet to a lavatory or bathtub shall be 15 inches (380 mm) minimum. The water closet shall be positioned to allow for future installation of a grab bar on the side with 18 inches (455 mm) clearance. Clearance around the water closet shall comply with Section 1004.11.3.1.2.1, 1004.11.3.1.2.2, or 1004.11.3.1.2.3.~~

~~**1004.11.3.1.2.1 Parallel Approach.** A clearance 56 inches (1420 mm) minimum measured from the wall behind the water closet, and 48 inches (1220 mm) minimum measured from a point 18 inches (455 mm) from the centerline of the water closet on the side designated for future installation of grab bars shall be provided. Vanities or lavatories on the wall behind the water closet are permitted to overlap the clearance.~~

~~**1004.11.3.1.2.2 Forward Approach.** A clearance 66 inches (1675 mm) minimum measured from the wall behind the water closet, and 48 inches (1220 mm) minimum measured from a point 18 inches (455 mm) from the centerline of the water closet on the side designated for future installation of grab bars shall be provided. Vanities or lavatories on the wall behind the water closet are permitted to overlap the clearance.~~

~~**1004.11.3.1.2.3 Parallel or Forward Approach.** A clearance 56 inches (1420 mm) minimum measured from the wall behind the water closet, and 42 inches (1065 mm) minimum measured from the centerline of the water closet shall be provided.~~

1004.11.3.1.2 Water Closet. The water closet shall comply with Section 1004.11.3.1.2.

1004.11.3.1.2.1 Location. The centerline of the water closet shall be 16 inches (405 mm) minimum and 18 inches (455 mm) maximum from one side of the required clearance.

1004.11.3.1.2.2 Clearance. Clearance around the water closet shall comply with Sections 1004.11.3.1.2.2.1 through 1004.11.3.1.2.2.3.

1004.11.3.1.2 continues

1004.11.3.1.2
Water Closets in Type B Units

1004.11.3.1.2 continued

Exception: Clearance complying with Sections 1003.11.2.4.2 through 1003.11.2.4.4.

1004.11.3.1.2.2.1 Clearance Width. Clearance around the water closet shall be 48 inches (1220 mm) minimum in width, measured perpendicular from the side of the clearance that is 16 inches (405 mm) minimum and 18 inches (455 mm) maximum from the water closet centerline.

1004.11.3.1.2.2.2 Clearance Depth. Clearance around the water closet shall be 56 inches (1420 mm) minimum in depth, measured perpendicular from the rear wall.

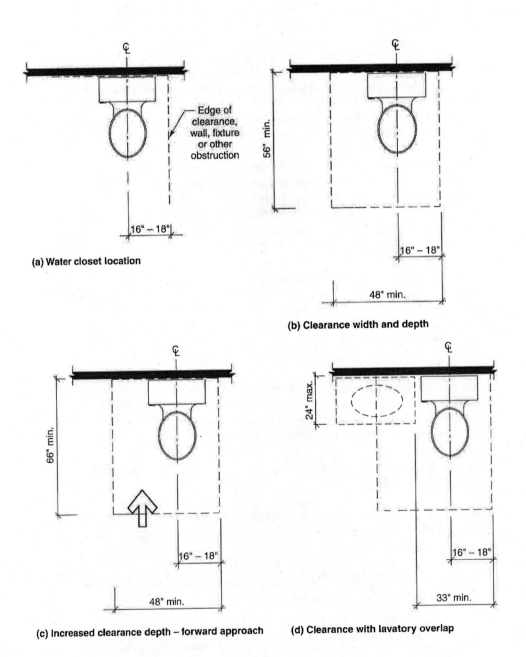

(a) Water closet location

(b) Clearance width and depth

(c) Increased clearance depth – forward approach

(d) Clearance with lavatory overlap

1004.11.3.1.2.2.3 Increased Clearance Depth at Forward Approach.

Where only a forward approach is provided, the clearance shall be 66 inches (1675 mm) minimum in depth, measured perpendicular from the rear wall.

1004.11.3.1.2.2.4 Clearance Overlap.

A vanity or other obstruction 24 inches (610 mm) maximum in depth, measured perpendicular from the rear wall, shall be permitted to overlap the required clearance, provided the width of the remaining clearance at the water closet is 33 inches (840 mm) minimum.

CHANGE SIGNIFICANCE: This change is essentially an editorial rewrite of the previous requirements. Whereas the standard previously provided a general section and then addressed the clearances based on the direction of approach, the new provisions are more concise and address the requirements by focusing on the clearances. The revision also allows the Type B units to more closely follow the format for Type A units and that of Chapter 6.

The main technical change is found in the location provisions of Section 1004.11.3.1.2.1, where a range of distance is now provided. Previously the water closets in a Type B unit required a distance of 18 inches from the centerline of the water closet to a wall or to a bathtub or lavatory opposite the direction of approach. The new requirements will generally change that 18-inch distance to a range of 16 inches minimum to 18 inches maximum. This range matches the requirements from Chapter 6 and the provisions for Accessible and Type A units and helps provide a tolerance for situations where the water closet is not located exactly at the 18-inch dimension from a wall. The 18-inch minimum will still be applicable if the water closet has a fixture on each side—a common configuration in a Type B unit, Option A bathroom.

Another smaller technical change can be found in the clearance overlap requirements of Section 1004.11.3.1.2.2.4. By revising the text from the previously permitted "vanities or lavatories" (2003 Sections 1004.11.3.1.2.1 and 1004.11.3.1.2.2) to a more general "vanity or other obstruction," it will allow a short wall or any other type of obstruction that sticks out the same depth as a lavatory or vanity. The provisions of the FHA allow a wall at this location, so it is nice that the standard will coordinate with the federal law and permit these types of obstructions. In order to limit the size of the obstruction, a 24-inch maximum depth was included that will match the depth of the typical vanity and the FHA provisions. The 33-inch clearance that is specified in this section is similar to the fact that the previous standard required a clearance of 18 inches on one side of the water closet and 15 inches on the other. So, while the dimension appears to be different, the resulting layout will be the same.

With the new sections each focusing on a specific clearance, versus addressing several issues, a large amount of redundant text was eliminated. For example, the 48-inch width requirement and the vanity overlap provisions were previously in both the parallel- and forward-approach sections and were identical. The new format places the 48-inch width requirement at a single location in Section 1004.11.3.1.2.2.1, while the vanity issue is in Section 1004.11.3.1.2.2.4.

1004.11.3.1.2 continues

1004.11.3.1.2 continued   When first looking at the requirements it may appear that the provisions related to the parallel- or forward-approach option from the old Section 1004.11.3.1.2.3 are missing. These requirements can be found by looking at the exception in Section 1004.11.3.1.2.2, which references the Type A unit requirements in Section 1003.11.2.4. By using the referenced provisions of Section 1003, all three of the previous approach options still exist and are permitted for the Type B units.

CHANGE TYPE: Addition

CHANGE SUMMARY: Provides technical requirements for dwelling units that are not regulated by the FHA, such as single-family homes and townhouses. The intent is so that the homes will both accommodate visitors with disabilities and promote the aging-in-place concept so that people may enjoy their homes without requiring major modifications when they, family, or friends face short- or long-term mobility constraints.

1005
Type C (Visitable) Units

2009 STANDARD:

1005 Type C (Visitable) Units

1005.1 General. Type C (Visitable) dwelling units shall comply with Section 1005.

1005.2 Unit Entrance. At least one unit entrance shall be on a circulation path complying with Section 1005.5 from a public street or sidewalk, a dwelling unit driveway, or a garage.

1005.3 Connected Spaces. A circulation path complying with Section 1005.5 shall connect the unit entrance complying with Section 1005.2 and with the spaces specified in Section 1005.4.

1005.4 Interior Spaces. The entrance level shall include a toilet room or bathroom complying with Section 1005.6 and one habitable space with an area 70 square feet minimum. Where a food preparation area is provided on the entrance level, it shall comply with Section 1005.7.

Exception: A toilet room or bathroom shall not be required on an entrance level with less than 120 square feet of habitable space.

1005.5 Circulation Path. Circulation paths shall comply with Section 1005.5.

1005.5.1 Components. The circulation path shall consist of one or more of the following elements: walking surfaces with a slope not steeper than 1:20, doors and doorways, ramps, elevators complying with Sections 407 through 409, and wheelchair (platform) lifts complying with Section 410.

1005.5.2 Walking Surfaces. Walking surfaces with slopes not steeper than 1:20 shall comply with Section 303.

1005.5.2.1 Clear Width. The clear width of the circulation path shall comply with Section 403.5.

1005.5.3 Doors and Doorways. Doors and doorways shall comply with Section 1005.5.3.

1005 continues

146

1005.5.3.1 Clear Width. Doorways shall have a clear opening of 31-3/4 inches (810 mm) minimum. Clear opening of swinging doors shall be measured between the face of the door and stop, with the door open 90 degrees.

1005.5.3.2 Thresholds. Thresholds shall comply with Section 303.

Exception: Thresholds at exterior sliding doors shall be permitted to be 3/4 inch (19 mm) maximum in height, provided they are beveled with a slope not steeper than 1:2.

1005.5.4 Ramps. Ramps shall comply with Section 405.

Exception: Handrails, intermediate landings, and edge protection are not required where the sides of ramp runs have a vertical drop off of ½ inch (13 mm) maximum within 10 inches (255 mm) horizontally of the ramp run.

1005.5.4.1 Clear Width. The clear width of the circulation path shall comply with Section 403.5.

1005.6 Toilet Room or Bathroom. At a minimum, the toilet room or bathroom required by Section 1005.4 shall include a lavatory and a water closet. Reinforcement shall be provided for the future installation of grab bars at water closets. Clearances at the water closet shall comply with Section 1004.11.3.1.2.

1005.7 Food Preparation Area. At a minimum, the food preparation area shall include a sink, a cooking appliance, and a refrigerator. Clearances between all opposing base cabinets, countertops, appliances, or walls within the food preparation area shall be 40 inches (1015 mm) minimum in width.

Exception: Spaces that do not provide a cooktop or conventional range shall be permitted to provide a clearance of 36 inches (915 mm) minimum in width.

1005.8 Lighting Controls and Receptacle Outlets. Receptacle outlets and operable parts of lighting controls shall be located 15 inches minimum and 48 inches maximum above the floor.

Exception: The following shall not be required to comply with Section 1005.8.

1. Receptacle outlets serving a dedicated use
2. Controls mounted on ceiling fans and ceiling lights
3. Floor receptacle outlets
4. Lighting controls and receptacle outlets over countertops

CHANGE SIGNIFICANCE: The purpose of this new section is to include the technical criteria that the A117 committee felt would be necessary to provide a level of accessibility for dwelling units that are visitable or will allow for "aging-in-place" features. The jurisdiction will need to specify when visitable units would be required (see Sections 201 and 202). A number of communities across the country have begun to develop or require homes to provide some basic level of accessibility. As these types of developments or rules have been adopted, each community had to create its own technical requirements. These local ordinances have typically required zero-step entrances, wider interior doors, and a few additional access features, but the requirements varied. By providing the Type C provisions in the standard, the committee has taken a step to provide consistency throughout the country and a model of what they believe is a minimum level of accessibility to accomplish the purpose of visitable or inclusive design.

Providing these minimum levels of accessibility will lead to dwelling units being constructed so that people who use mobility devices or have difficulty climbing stairs are able to continue to live in their own homes or visit the homes of others. One of the difficulties of establishing this section was determining exactly what level of access was needed and appropriate for a minimal level of access. Primarily the requirements will establish that portions of a dwelling unit are accessed by a circulation path that does not include any stairways or abrupt level changes. Once inside, the interior path should connect to a toilet room, a habitable space, and if provided on the entry level, a food preparation area. The circulation path matches the accessible route's width provisions of Section 403.5 but would need to be wider in certain food preparation areas. The last required element of accessibility is that most lighting controls and receptacle outlets must be located within the specified reach range.

If these few access features are provided, people with mobility impairments can visit the homes of friends or family. Furthermore, these features may permit residents who are injured, develop a disability, or are recovering from an operation to remain living in their homes for a short time, even if lacking a full bathroom and a separate designated sleeping

1005 continues

1005 continued

space, while they plan and make any additional renovations they might need or while seeking a different place to reside.

The standard does not address the scoping requirements for when and in what quantity the Type C (visitable) units would be required. This type of scoping would be developed by the jurisdiction as specified in Sections 201 and 202. During the A117.1 standard development process it was the committee's assumption that these units would generally be applied to dwellings that were not regulated by the requirements of the FHA. Therefore it would typically be applied to structures with three or fewer dwelling units within them.

To quickly address some of the main features, the following comments are provided.

- Section 1005.2: The provision requires at least one entrance to be accessible, but unlike the other three types of units, this section does not mandate that the primary entry be selected or limit the path within the unit. In lieu of the primary entry, other viable options would be the door into the home from the garage or a door off a back patio.

- Section 1005.3: The route provided into and through the dwelling should meet most provisions of accessible routes but, since it does not meet all requirements, it should not be called an "accessible route." To avoid confusion with other requirements of the standard, the term "circulation path" is used within the requirements.

- Section 1005.5.4: The standard provides an exception to eliminate handrails or edge protection on ramps that move up with the surrounding grade. Residents of single-family homes often prefer not to have obstructions such as handrails and edge protection along their walkways where they can prevent circulation to portions of the yard or appear different from neighboring homes. The committee believes that handrails and edge protection can be added by an owner when and if needed, but does not believe that every home subject to the modest requirements for Type C units should incur the added cost associated with such elements.

- Section 1005.6: A toilet room or bathroom must be provided that allows for clearances and blocking at the water closet. Clearances are not indicated for the lavatory or any bathing facilities that are provided. Providing clearances at these elements would be best design practice, but clearances are not specifically required.

- Section 1005.7: Although visitability requirements or policies do not always specify access to kitchens and eating areas, they have been included in the standard to address the goals of the Type C unit. The food preparation area may, however, simply be an auxiliary kitchen such as a bar area with a small refrigerator and microwave. It is important to note that this requirement is scoped by Section 1005.4 and would only be applicable where a food preparation area was provided on the entrance level.

- Section 1005.8: The requirement for electrical outlets and light switches to be within reach range heights is a limited application of operable parts requirements found in Accessible, Type A, and Type B units. There are no requirements for other operable parts, such as appliances, plumbing fixtures, door hardware, and so on.

CHANGE TYPE: Addition

CHANGE SUMMARY: Adds a new chapter into the standard to provide the technical requirements for making sports and recreational facilities accessible. Includes provisions for items such as amusement rides, boating facilities and fishing piers, golf, playground equipment, swimming pools, and more.

2009 STANDARD:

See the standard for complete text. Because this is an entirely new section, the complete text is not shown here.

Chapter 11
Recreational Facilities

Chapter 11. Recreational Facilities

1101 General

1101.1 Scope. Recreational facilities required to be accessible by the scoping provisions adopted by the administrative authority shall comply with the applicable provisions of Chapter 11.

1101.2 Special Provisions

1101.2.1 General Exceptions. The following shall not be required to be accessible or to be on an accessible route:

1. Raised structures used solely for refereeing, judging, or scoring a sport
2. Water slides
3. Animal containment areas that are not for public use
4. Raised boxing or wrestling rings
5. Raised diving boards and diving platforms
6. Bowling lanes that are not required to provide wheelchair spaces
7. Mobile or portable amusement rides
8. Amusement rides that are controlled or operated by the rider
9. Amusement rides designed primarily for children, where children are assisted on and off the ride by an adult
10. Amusement rides that do not provide amusement ride seats

1101.2.2 Area of Sport Activity. Areas of sport activity shall be served by an accessible route and shall not be required to be accessible except as provided in Chapter 11.

1101.2.3 Recreational Boating Facilities. Operable parts of cleats and other boat securement devices shall not be required to comply with Section 308.

Chapter 11 continues

Chapter 11 continued

1101.2.4 Exercise Machines and Equipment. Exercise machines and exercise equipment shall not be required to comply with Section 309.

1101.3 Protruding Objects. Protruding objects on circulation paths shall comply with Section 307.

Exceptions:

1. Within areas of sport activity, protruding objects on circulation paths shall not be required to comply with Section 307.

2. Within play areas, protruding objects on circulation paths shall not be required to comply with Section 307 provided that ground level accessible routes provide vertical clearance complying with Section 1108.2.

1102 Amusement Rides

1103 Recreational Boating Facilities

1104 Exercise Machines and Equipment

1105 Fishing Piers and Platforms

1106 Golf Facilities

1107 Miniature Golf Facilities

1108 Play Areas

1109 Swimming Pools, Wading Pools, Hot Tubs, and Spas

1110 Shooting Facilities with Firing Positions

CHANGE SIGNIFICANCE: Specific technical requirements for making sports and recreational facilities accessible were not previously included in the standard. While the IBC and other scoping documents that reference the A117.1 standard require certain elements on a site to be on accessible routes, such as a swimming pool or tennis court at an apartment complex, the standard did not address access for the element or facility once the route reached it. This new chapter will provide the specifications for the element so that certain portions or aspects are accessible. The inclusion of these new requirements and the scoping that will be placed in the IBC show that the accessibility requirements are taking a different approach and are branching out into providing access from the perspective of participation in the recreation, rather than just watching.

A number of new definitions have been added to Chapter 1 of the standard as a part of adding this chapter. A review of the definitions is appropriate to understand the requirements of Chapter 11. Definitions have been added for the types or components of recreational facilities, such as "area of sport activity," "catch pool," "play area," and so on. Safety standards are referenced in Chapter 1 for playground equipment and the surfaces around the playground equipment.

The requirements of the standard and the scoping from the IBC will not require the provisions of accessible routes for every player position on a playing field. The primary idea is that someone can get to a recreational facility, and participation in that event will depend on his or her own abilities. The requirements that are included in Chapter 11 are developed from the ADA and ABA AG and will help harmonize the standard with the requirements of the federal law. These requirements were developed through the work of the U.S. Access Board with the technical requirements and scoping being placed in Chapter 10 and Chapter 2 of the ADA

Chapter 11 continues

Chapter 11 continued

and ABA AG. Although the A117.1 standard generally does not contain scoping (see Sections 201 and 202), the recreational facilities requirements of Chapter 11 do provide general exemptions for certain elements that are not required to be accessible or to be on an accessible route. Until scoping for these elements is provided in the IBC, or is generally accepted, including the exemptions provides better guidance regarding the intent and would help make enforcement more consistent by clarifying what portions of the recreational facilities the committee feels do not need to be or may be too difficult to make accessible. Including the exemptions or scoping in the standard will also help harmonize with the federal law.

As this book is being developed, there is a code-change proposal that has been submitted to the ICC that would add scoping for recreational facilities into the 2012 edition of the IBC. Information related to the IBC code-development process can be found on the ICC website, http://www.iccsafe.org.

1102
Amusement Rides

CHANGE TYPE: Addition

CHANGE SUMMARY: This new section provides technical criteria for amusement rides that move a person through a fixed course for the purpose of amusement. This could include rides such as cars, boats, coasters, and so forth that the rider would either board or transfer into. There are several exemptions for different types of amusement rides in Section 1101.2.1.

2009 STANDARD:

See the standard for complete text. Because this is an entirely new section, the complete text is not shown here.

amusement ride. A system that moves persons through a fixed course within a defined area for the purpose of amusement.

1102 Amusement Rides

1102.1 General. Accessible Amusement rides shall comply with Section 1102.

1102.2 Accessible Routes. Accessible routes serving amusement rides shall comply with Chapter 4.

Exceptions:

1. In load or unload areas and on amusement rides, where complying with Section 405.2 is not structurally or operationally feasible, ramp slope shall be permitted to be 1:8 maximum.
2. In load or unload areas and on amusement rides, handrails provided along walking surfaces complying with Section 403 and required on ramps complying with Section 405 shall not be required to comply with Section 505 where complying is not structurally or operationally feasible.

1102.3 Load and Unload Areas. A turning space complying with Sections 304.2 and 304.3 shall be provided in load and unload areas.

1102.4 Wheelchair Spaces in Amusement Rides. Wheelchair spaces in amusement rides shall comply with Section 1102.4.

1102.4.1 Floor Surface.

1102.4.2 Slope.

1102.4.3 Gaps. Floors of amusement rides with wheelchair spaces and floors of load and unload areas shall be coordinated so that, when amusement rides are at rest in the load and unload position, the vertical difference between the floors shall be within plus or minus 5/8 inches (16 mm) and the horizontal gap shall be 3 inches (75 mm) maximum under normal passenger load conditions.

1102 continues

154

1102 continued

Courtesy U.S. Access Board

Exception: Where complying is not operationally or structurally feasible, ramps, bridge plates, or similar devices complying with the applicable requirements of 36 CFR 1192.83(c) listed in Section 105.2.11 shall be provided.

1102.4.4 Clearances. Clearances for wheelchair spaces shall comply with Section 1102.4.4.

Exceptions:

1. Where provided, securement devices shall be permitted to overlap required clearances.
2. Wheelchair spaces shall be permitted to be mechanically or manually repositioned.
3. Wheelchair spaces shall not be required to comply with Section 307.4.

1102.4.4.1 Width and Length.

1102.4.4.2 Side Entry.

1102.4.4.3 Permitted Protrusions in Wheelchair Spaces. Objects are permitted to protrude a distance of 6 inches (150 mm) maximum along the front of the wheelchair space, where located 9 inches (230 mm) minimum and 27 inches (685 mm) maximum above the floor of the wheelchair space. Objects are permitted to protrude a distance of 25 inches (635 mm) maximum along the front of the wheelchair space, where located more than 27 inches (685 mm) above the floor of the wheelchair space.

1102.4.5 Ride Entry.

1102.4.6 Approach. One side of the wheelchair space shall adjoin an accessible route when in the load and unload position.

1102.4.7 Companion Seats.

1102.4.7.1 Shoulder-to-Shoulder Seating.

1102.5 Amusement Ride Seats Designed for Transfer. Amusement ride seats designed for transfer shall comply with Section 1102.5 when positioned for loading and unloading.

1102.5.1 Clear Floor Space.

1102.5.2 Transfer Height.

1102.5.3 Transfer Entry.

1102.5.4 Wheelchair Storage Space. Wheelchair storage spaces complying with Section 305 shall be provided in or adjacent to unload areas for each required amusement ride seat designed for transfer and shall not overlap any required means of egress or accessible route.

1102.6 Transfer Devices for Use with Amusement Rides. Transfer devices for use with amusement rides shall comply with Section 1102.6 when positioned for loading and unloading.

1102.6.1 Clear Floor Space.

1102 continues

1102 continued   **1102.6.2 Transfer Height.**

1102.6.3 Wheelchair Storage Space. Wheelchair storage spaces complying with Section 305 shall be provided in or adjacent to unload areas for each required transfer device and shall not overlap any required means of egress or accessible route.

CHANGE SIGNIFICANCE: The addition of this new section helps provide technical criteria for amusement rides as a part of the new Chapter 11. Including these provisions within the standard will help coordinate with the ADA and ABA AG. It is important to distinguish that the requirements of Sections 1102.1, 1102.2, and 1102.3 will apply to the building or facility that serves the amusement ride, while the remaining portions of the section (Sections 1102.1, 1102.4, 1102.5, and 1102.6) apply to the ride itself. Therefore code officials will generally focus only on the building portions, while ride manufacturers would be concerned with the ride and the later sections. Knowing what is being regulated is important when applying the requirements. For example, the protruding object exception (Exception 3) and the other exceptions in Section 1102.4.4 along with Section 1102.4.4.3 apply to the circulation space and clear space requirements on the ride. These exceptions do not apply to the circulation areas and accessible routes in the queue line or in the load and unload areas.

A few of the items that may be helpful when applying these requirements are:

- Users should review the new definitions in Chapter 1 for several terms that affect the application of these sections.
- Section 1102.2: Because the exceptions may be used in new construction, the term "structurally or operationally infeasible" is used in lieu of technically infeasible. The provisions do not intend to negate a manufacturer's warranty by requiring a modification that may change a ride's structural or operational characteristics to the extent that the ride's performance differs from that specified by the manufacturers. Additionally, the exceptions are limited to the load and unload areas and generally are only used where these areas connect to the ride. The exceptions do not apply for other portions of the accessible route, such as the queue line leading to the load and unload areas.
- Section 1102.4.3, Exception: The reference to (Code of Federal Regulations) CFR 1192.83 (c) is to the U.S. Access Board's vehicle accessibility guidelines, which allow several different manually deployed devices to be used in the transition from the load and unload space onto the ride. The technical requirements for the ramps and bridge plates are based on the provisions for light rail vehicles.
- Sections 1102.5.4 and 1102.6.3: A clear space is needed in the load and unload areas for individuals to leave their wheelchairs when they transfer onto amusement rides. The minimum size of the space is 30 inches by 48 inches based on the reference and Section 305.3. For obvious safety reasons, the space must not overlap any required means of egress or accessible route. This provision does not require a constructed element for storage, only a space. Although these are technically a ride requirement due to the location in Sections 1102.5 and 1102.6, designers should be aware of this provision when designing the building or facility that serves the ride.
- Additional information and guidance can be found within the ADA and ABA AG on the U.S. Access Board website, http://www.access-board.gov.

CHANGE TYPE: Addition

CHANGE SUMMARY: Provisions have been added to address boating facilities and also fishing piers. One of the key issues addressed is the dynamic interface between the land and the water for items such as floating facilities.

1103, 1105
Recreational Boating Facilities and Fishing Piers and Platforms

2009 STANDARD:

See the standard for complete text. Because these are lengthy new sections, the complete text is not shown here.

boarding pier. A portion of a pier where a boat is temporarily secured for the purpose of embarking or disembarking.

boat launch ramp. A sloped surface designed for launching and retrieving trailered boats and other water craft to and from a body of water.

boat slip. That portion of a pier, main pier, finger pier, or float where a boat is moored for the purpose of berthing, embarking, or disembarking.

gangway. A variable-sloped pedestrian walkway that links a fixed structure or land with a floating structure. Gangways that connect to vessels are not addressed by this document.

1103 Recreational Boating Facilities

1103.1 General. Accessible recreational boating facilities shall comply with Section 1103.

1103.2 Accessible Routes. Accessible routes serving recreational boating facilities, including gangways and floating piers, shall comply with Chapter 4 except as modified by the exceptions in Section 1103.2.

1103.2.1 Boat Slips. An accessible route shall serve boat slips.

Exceptions: (Nine exceptions)

1103.2.2 Boarding Piers at Boat Launch Ramps. An accessible route shall serve boarding piers.

Exceptions: (Three exceptions)

1103.3 Clearances. Clearances at boat slips and on boarding piers at boat launch ramps shall comply with Section 1103.3.

1103.3.1 Boat Slip Clearance. Boat slips shall provide clear pier space 60 inches (1525 mm) minimum in width that extend the full length of the boat slips. Each 10 feet (3050 mm) of linear pier edge serving boat slips shall contain at least one continuous clear opening 60 inches (1525 mm) minimum in width.

1103, 1105 continues

158

1103, 1105 continued

Exceptions: (Three exceptions)

1103.3.2 Boarding Pier Clearances. Boarding piers at boat launch ramps shall provide clear pier space 60 inches (1525 mm) minimum in width and shall extend the full length of the boarding pier. Every 10 feet (3050 mm) of linear pier edge shall contain at least one continuous clear opening 60 inches (1525 mm) minimum in width.

Exceptions: (Two exceptions)

1105 Fishing Piers and Platforms

1105.1 Accessible Routes. Accessible routes serving fishing piers and platforms, including gangways and floating piers, shall comply with Chapter 4.

Exceptions:

1. Accessible routes serving floating fishing piers and platforms shall be permitted to use Exceptions 1, 2, 5, 6, 7, and 8 in Section 1103.2.1.
2. Where the total length of the gangway or series of gangways serving as part of a required accessible route is 30 feet (9145 mm) minimum, gangways shall not be required to comply with Section 405.2.

1105.2 Railings. Where provided, railings, guards, or handrails shall comply with Section 1105.2.

1105.2.1 Height. A minimum of 25 percent of the railings, guards, or handrails shall be 34 inches (865 mm) maximum above the ground or deck surface.

Exception: Where a guard complying with the applicable building code is provided, the guard shall not be required to comply with Section 1105.2.1.

1105.2.1.1 Dispersion. Railings, guards, or handrails required to comply with Section 1105.2.1 shall be dispersed throughout the fishing pier or platform.

1105.3 Edge Protection. Where railings, guards, or handrails complying with Section 1105.2 are provided, edge protection complying with Sections 1105.3.1 or 1105.3.2 shall be provided.

1105.3.1 Curb or Barrier. Curbs or barriers shall extend 2 inches (51 mm) minimum in height above the surface of the fishing pier or platform.

1105.3.2 Extended Ground or Deck Surface. The ground or deck surface shall extend 12 inches (305 mm) minimum beyond the inside face of the railing. Toe clearance shall be provided and shall be 30 inches (760 mm) minimum in width and 9 inches (230 mm) minimum in height above the ground or deck surface beyond the railing.

1105.4 Clear Floor Space. At each location where there are railings, guards, or handrails complying with Section 1105.2.1, a clear floor space complying with Section 305 shall be provided. Where there are no railings, guards, or handrails, at least one clear floor space complying with Section 305 shall be provided on the fishing pier or platform.

1105.5 Turning Space. At least one turning space complying with Section 304.3 shall be provided on fishing piers and platforms.

CHANGE SIGNIFICANCE: These two new sections provide the technical criteria for making recreational boating facilities as well as fishing piers and platforms accessible. In regards to the boating facility requirements, it is important to realize that the provisions are not intended to apply to passenger vessel facilities but only to recreational boating facilities.

Gangways and floating piers that serve recreational boating facilities must be able to serve as part of an accessible route. The dynamic interface between land and water presents unique and significant challenges in providing access to floating facilities. The key challenge is the vertical variations caused by changes in water levels due to tides and seasonal water levels. The criteria for gangways take this into account. There are exceptions for maximum rise, slope, handrail extensions, and level landings. Sloped accessible gangways are not required to be longer than 80 feet. The provisions of this section recognize the challenges the changing water level can create, and therefore the provisions provide an exception to the general requirements of Sections 303 and 405 related to changes in levels and ramps. There are specific requirements for clearances at the accessible boating slips.

The 60-inch minimum width that is required at several of the sections is consistent with the width required at the access aisle for the standard accessible parking spaces. However, unlike vehicle access aisles, it is not necessary for the entire pier to have a 60-inch clear width. This is done in recognition that obstructions are allowed to be located around the edge of the piers and could be adjacent to the 60-inch minimum openings at the pier edge.

1103, 1105 continues

160

1103, 1105 continued

The route to fishing piers and platforms is handled the same as boating facilities. When guards are provided, a portion must be positioned at a height of 34 inches or lower to allow for someone using a wheelchair to be able to fish over the rail. This lower guard height may be considered as being conceptually similar to the reduced guard height that is permitted in front of assembly seating spaces. This is in recognition that a full-height guard would obstruct the intended purpose of allowing fishing over the railing. Where guards are required for safety, the lowered portions are not required. Edge protection of at least 2 inches in height must be provided to prevent the wheels of mobility aids from slipping over the edge.

Devices to secure the boats to the pier are exempted from the reach range requirements by Section 1101.2.3.

1106, 1107

Golf Facilities and Miniature Golf Facilities

CHANGE TYPE: Addition

CHANGE SUMMARY: An accessible route and other features are required to be provided for portions of a general golf course and also for a miniature golf course. Some of these provisions modify the general accessible route requirements. Portions of the golf course requirements are established so that a golf cart could be used for access rather than other mobility devices.

2009 STANDARD:

golf car passage. A continuous passage on which a motorized golf car can operate.

teeing ground. In golf, the starting place for the hole to be played.

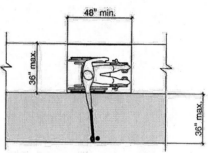

Note: Running slope of clear floor or ground space not steeper than 1:20

1106 Golf Facilities

1106.1 General. Golf facilities shall comply with Section 1106.

1106.2 Accessible Routes. Accessible routes serving teeing grounds, practice teeing grounds, putting greens, practice putting greens, teeing stations at driving ranges, course weather shelters, golf cart rental areas, bag drop areas, and course toilet rooms shall comply with Chapter 4 and shall be 48 inches (1220 mm) minimum in width. Where handrails are provided, accessible routes shall be 60 inches (1525 mm) minimum in width.

Exception: Handrails shall not be required on golf courses. Where handrails are provided on golf courses, the handrails shall not be required to comply with Section 505.

1106.3 Golf Car Passages. Golf car passages shall comply with Section 1106.3.

1106.3.1 Clear Width. The clear width of golf car passages shall be 48 inches (1220 mm) minimum.

1106.3.2 Barriers. Where curbs or other constructed barriers prevent golf cars from entering a fairway, openings 60 inches (1525 mm) minimum in width shall be provided at intervals not to exceed 75 yards (69 m).

1106.4 Weather Shelters. A clear floor space 60 inches (1525 mm) minimum by 96 inches (2440 mm) minimum shall be provided within weather shelters.

1107 Miniature Golf Facilities

1107.1 General. Miniature golf facilities shall comply with Section 1107.

1106, 1107 continues

1106, 1107 continued

1107.2 Accessible Routes. Accessible routes serving holes on miniature golf courses shall comply with Chapter 4.

Exception: Accessible routes located on playing surfaces of miniature golf holes shall be permitted to comply with the following:

1. Playing surfaces shall not be required to comply with Section 302.2.
2. Where accessible routes intersect playing surfaces of holes, a curb that is 1 inch (25 mm) maximum in height and 32 inches (815 mm) minimum in width shall be permitted.
3. A slope of 1:4 maximum shall be permitted for a rise of 4 inches (100 mm) maximum.
4. Ramp landing slopes specified by Section 405.7.1 shall be permitted to be 1:20 maximum.
5. Ramp landing length specified by Section 405.7.3 shall be permitted to be 48 inches (1220 mm) in length minimum.
6. Ramp landing size at a change in direction specified by Section 405.7.4 shall be permitted to be 48 inches (1220 mm) minimum by 60 inches (1525 mm) minimum.
7. Handrails shall not be required on holes. Where handrails are provided on holes, the handrails shall not be required to comply with Section 505.

1107.3 Miniature Golf Holes. Miniature golf holes shall comply with Section 1107.3.

1107.3.1 Start of Play. A clear floor space 48 inches (1220 mm) minimum by 60 inches (1525 mm) minimum with slopes not steeper than 1:48 shall be provided at the start of play.

1107.3.2 Golf Club Reach Range Area. All areas within holes where golf balls rest shall be within 36 inches (915 mm) maximum of a clear floor space 36 inches (915 mm) minimum in width and 48 inches (1220 mm) minimum in length having a running slope not steeper than 1:20. The clear floor space shall be served by an accessible route.

CHANGE SIGNIFICANCE: The requirements of these two sections are intended to provide the criteria to make both regular golf courses and miniature golf facilities accessible. These two sections are a good example of how the provisions are moving to focus on the participation of the users and not simply providing access to view or observe the game.

Access to golf courses is typically achieved through the use of golf carts. These carts are permitted as an alternative to providing accessible routes throughout golf courses. Courses must be designed so the golf carts can access teeing grounds and putting greens. Accessible routes are required to other areas such as practice greens, driving ranges, cart rental areas, and bathrooms.

The 48-inch minimum width that is required for an accessible route in Section 1106.2 is necessary to ensure passage of a golf car on either the accessible route or the golf car passage. This route requirement is important where the accessible route is used to connect the golf car rental area, bag drop areas, practice putting greens, accessible practice teeing grounds, course toilet rooms, and course weather shelters. These areas are considered as being outside the boundary of the golf course, but are areas where an individual using an adapted golf car may travel. A golf car passage may not be substituted for other accessible routes located outside the boundary of the course.

At least half of the holes in a miniature golf course must be on an accessible route. However, the requirements allow for miniature golf design conventions such as curbs. All level areas of an accessible hole where a ball may come to rest must be within 36 inches of the route so that someone can reach the ball with a club.

Exception 3 in Section 1107.2 is only applicable where the accessible route connecting the start of play areas is located on the hole. Permitting a greater slope for this limited distance allows more flexibility in the design of the hole while still providing access on the hole for the play of the game. Exception 7 of the same section eliminates the requirement for compliant handrails on a ramp located on a miniature golf hole. When the U.S. Access Board was originally developing these requirements, designers and operators of miniature golf facilities cautioned against requiring the handrails because of the potential danger of golf balls ricocheting off of them. Notice, however, that this exception is only applicable on the hole and not to the general circulation route within the facility.

1108

Play Areas

CHANGE TYPE: Addition

CHANGE SUMMARY: This section is a continuation of the coordination effort of the A117.1 standard and the ADA and ABA AG. This section addresses a variety of recreational play areas.

2009 STANDARD:

See the standard for complete text. Because this is an entirely new section, the complete text is not shown here.

elevated play component: A play component that is approached above or below grade and that is part of a composite play structure consisting of two or more play components attached or functionally linked to create an integrated unit providing more than one play activity.

ground level play component: A play component that is approached and exited at the ground level.

play area: A portion of a site containing play components designed and constructed for children.

play component: An element intended to generate specific opportunities for play, socialization, or learning. Play components are manufactured or natural; and are stand-alone or part of a composite play structure.

soft contained play structure: A play structure made up of one or more play components where the user enters a fully enclosed play environment that utilizes pliable materials, such as plastic, netting, or fabric.

use zone: The ground level area beneath and immediately adjacent to a play structure or play equipment that is designated by ASTM F 1487 listed in Section 105.2.10 for unrestricted circulation around the play equipment and where it is predicted that a user would land when falling from or exiting the play equipment.

1108 Play Areas

1108.1 Scope. Play areas shall comply with 1108.

1108.2 Accessible Route for Play Areas. Play areas shall provide accessible routes in accordance with Section 1108.2. Accessible routes serving play areas shall comply with Chapter 4 except as modified by Section 1108.4.

1108.2.1 Ground Level and Elevated Play Components. At least one accessible route shall be provided within the play area. The accessible route shall connect ground level play components required to comply with Section 1108.3.2.1 and elevated play components required to comply with Section 1108.3.2.2, including entry and exit points of the play components.

1108.2.2 Soft Contained Play Structures. Where three or fewer entry points are provided for soft contained play structures, at least one entry point shall be on an accessible route. Where four or more entry points are provided for soft contained play structures, at least two entry points shall be on an accessible route.

1108.3 Age Groups. Play areas for children ages 2 and over shall comply with Section 1108.3. Where separate play areas are provided within a site for specific age groups, each play area shall comply with Section 1108.3.

Exceptions:

1. Play areas located in family child care facilities where the proprietor actually resides shall not be required to comply with Section 1108.3.

2. In existing play areas, where play components are relocated for the purposes of creating safe use zones and the ground surface is not altered or extended for more than one use zone, the play area shall not be required to comply with Section 1108.3.

3. Amusement attractions shall not be required to comply with Section 1108.3.

4. Where play components are altered and the ground surface is not altered, the ground surface shall not be required to comply with Section 1108.4.1.6 unless required by the authority having jurisdiction.

1108.3.1 Additions.

1108 continues

166

CHANGE SIGNIFICANCE: This section addresses a variety of different play areas in order to provide access to them. The included provisions regulate ground-level and elevated play components. The requirements within the standard include sections dealing with self-contained play structures, water play components, play tables, and so forth.

Playground access is required in facilities such as those at schools, parks, and day care facilities (not including day care facilities in a private home). Portions of playground equipment must be on an accessible route and allow children to either play while still in their wheelchairs, or have a transfer surface so they can move up into the equipment. Requirements include provisions for the number of play components required to be accessible, accessible surfacing, ramp access, transfer systems to access elevated structures, and access to soft-contained play structures.

Section 1101.3 provides an exemption from the protruding object requirements for certain portions of the play area.

1109

Swimming Pools, Wading Pools, Hot Tubs, and Spas

CHANGE TYPE: Addition

CHANGE SUMMARY: This section provides the technical criteria for swimming pools, wading pools, hot tubs, and spas. While these items previously needed to be on an accessible route, these provisions include the number and means of entry points to make different types of pools accessible.

2009 STANDARD:

See the standard for complete text. Because this is an entirely new section, the complete text is not shown here.

catch pool. A pool or designated section of a pool used as a terminus for water slide flumes.

1109 Swimming Pools, Wading Pools, Hot Tubs, and Spas

1109.1 General. Swimming pools, wading pools, hot tubs, and spas shall comply with Section 1109.

1109.1.1 Swimming pools. At least two accessible means of entry shall be provided for swimming pools. Accessible means of entry shall be swimming pool lifts complying with Section 1109.2; sloped entries complying with Section 1109.3; transfer walls complying with Section 1109.4 transfer systems complying with Section 1109.5; and pool stairs complying with Section 1109.6. At least one accessible means of entry provided shall comply with Sections 1109.2 or 1109.3.

Exceptions:

1. Where a swimming pool has less than 300 linear feet (91 m) of swimming pool wall, no more than one accessible means of entry shall be required.

2. Wave action pools, leisure rivers, sand bottom pools, and other pools where user access is limited to one area shall not be required to provide more than one accessible means of entry provided that the accessible means of entry is a swimming pool lift complying with Section 1109.2, a sloped entry complying with Section 1109.3, or a transfer system complying with Section 1109.5.

3. Catch pools shall not be required to provide an accessible means of entry provided that the catch pool edge is on an accessible route.

1109.1.2 Wading Pools. At least one sloped entry complying with Section 1109.3 shall be provided in wading pools.

1109.1.3 Hot Tubs and Spas. At least one accessible means of entry shall be provided for hot tubs and spas. Accessible means of entry shall comply with swimming pool lifts complying with Section 1109.2; transfer walls complying with Section 1109.4; or transfer systems complying with Section 1109.5.

Exception: Where hot tubs or spas are provided in a cluster, no more than 5 percent, but not less than one hot tub or spa in each cluster shall be required to comply with Section 1109.1.3.

1109.2 Pool Lifts.

1109.3 Sloped Entries. Sloped entries shall comply with Section 1109.3.

1109.3.1 Sloped Entry Route. Sloped entries shall comply with Chapter 4 except as modified in Sections 1109.3.1 through 1109.3.3.

Exception: Where sloped entries are provided, the surfaces shall not be required to be slip resistant.

1109.4 Transfer Walls. Transfer walls shall comply with Section 1109.4.

1109.4.1 Clear Deck Space. A clear deck space of 60 inches (1525 mm) minimum by 60 inches (1525 mm) minimum with a slope not steeper than 1:48 shall be provided at the base of the transfer wall. Where one grab bar is provided, the clear deck space shall be centered on the grab bar. Where two grab bars are provided, the clear deck space shall be centered on the clearance between the grab bars.

1109.4.2 Height.

1109.4.3 Wall Depth and Length. The transfer wall shall be 12 inches (305 mm) minimum and 16 inches (405 mm) maximum in depth. The transfer wall shall be 60 inches (1525 mm) minimum in length and shall be centered on the clear deck space.

1109 continues

1109 continued

1109.4.4 Surface.

1109.4.5 Grab Bars.
At least one grab bar complying with Sections 609.1 through 609.3 and 609.5 through 609.8 shall be provided on the transfer wall. Grab bars shall be perpendicular to the pool wall and shall extend the full depth of the transfer wall. The top of the gripping surface shall be 4 inches (100 mm) minimum and 6 inches (150 mm) maximum above the transfer wall. Where one grab bar is provided, clearance shall be 24 inches (610 mm) minimum on both sides of the grab bar. Where two grab bars are provided, clearance between grab bars shall be 24 inches (610 mm) minimum.

1109.5 Transfer Systems.

1109.6 Pool Stairs.
Pool stairs shall comply with Section 1109.6.

1109.6.1 Pool Stairs.
Pool stairs shall comply with Section 504.

Exception: Pool step risers shall not be required to be 4 inches (100 mm) minimum and 7 inches (180 mm) maximum in height provided that riser heights are uniform.

1109.6.2 Handrails.

CHANGE SIGNIFICANCE: The number of accessible entry points into a swimming pool is based on the size and type of the pool. All pools must have at least one accessible point of entry. Larger swimming pools (those with 300 or more linear feet of pool wall) must have two entry points. Options include pool lifts, sloped entries, transfer walls, transfer stairs, or ramp access. Allowances are made for wading pools, spas, wave action pools, and other types of pools where user access is limited to one area. Water slides, diving boards, and diving platforms are not required to be accessible based on Section 1101.2.1. The catch pools that serve them are not required to provide an accessible means of entry based on Exception 3 in 1109.1.1.

Although ramps are a typical means of providing access, they may not be the best solution for pool access. Because there are no requirements for pools to provide bathing chairs that can move into the water and due to possible entrapment issues associated with handrails in the pool, standard ramps may be less effective in providing pool access.

One of the important access requirements can be found in the last sentence of the base paragraph in Section 1109.1.1. That sentence will require that "at least one accessible means of entry" into the pool must be provided by either a pool lift or a sloped entry.

There are also special considerations for small areas such as hot tubs and spas within Section 1109.

Index

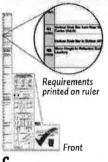
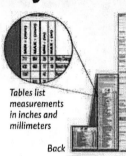

INTERNATIONAL CODE COUNCIL

People Helping People Build a Safer World™

Dedicated to the Support of Building Safety and Sustainability Professionals

An Overview of the International Code Council

The International Code Council (ICC) is a membership association dedicated to building safety, fire prevention and sustainability in the design and construction of residential and commercial buildings, including homes and schools. Most U.S. cities, counties, states and U.S. territories, and a growing list of international bodies, that adopt building safety codes use ones developed by the International Code Council.

Services of the ICC

The organizations that comprise the International Code Council offer unmatched technical, educational and informational products and services in support of the International Codes, with more than 250 highly qualified staff members at 16 offices throughout the United States, Latin America and the Middle East. Some of the products and services readily available to code users include:

- **CODE APPLICATION ASSISTANCE**
- **EDUCATIONAL PROGRAMS**
- **CERTIFICATION PROGRAMS**
- **TECHNICAL HANDBOOKS AND WORKBOOKS**
- **PLAN REVIEW SERVICES**
- **CODE COMPLIANCE EVALUATION SERVICES**
- **ELECTRONIC PRODUCTS**
- **MONTHLY ONLINE MAGAZINES AND NEWSLETTERS**

- **PUBLICATION OF PROPOSED CODE CHANGES**
- **TRAINING AND INFORMATIONAL VIDEOS**
- **BUILDING DEPARTMENT ACCREDITATION PROGRAMS**
- **GREEN BUILDING PRODUCTS AND SERVICES INCLUDING PRODUCT SUSTAINABILITY TESTING**

The ICC family of non-profit organizations include:

ICC EVALUATION SERVICE (ICC-ES)

ICC-ES is the United States' leader in evaluating building products for compliance with code. A nonprofit, public-benefit corporation, ICC-ES does technical evaluations of building products, components, methods, and materials.

ICC FOUNDATION (ICCF)

ICCF is dedicated to consumer education initiatives, professional development programs to support code officials and community service projects that result in safer, more sustainable buildings and homes.

INTERNATIONAL ACCREDITATION SERVICE (IAS)

IAS accredits testing and calibration laboratories, inspection agencies, building departments, fabricator inspection programs and IBC special inspection agencies.

NEED MORE INFORMATION? CONTACT ICC TODAY!
1-888-ICC-SAFE (422-7233)
www.iccsafe.org

10-03430

Don't Miss Out On Valuable ICC Membership Benefits. Join ICC Today!

Join the largest and most respected building code and safety organization. As an official member of the International Code Council®, these great ICC® benefits are at your fingertips.

EXCLUSIVE MEMBER DISCOUNTS

ICC members enjoy exclusive discounts on codes, technical publications, seminars, plan reviews, educational materials, videos, and other products and services.

TECHNICAL SUPPORT

ICC members get expert code support services, opinions, and technical assistance from experienced engineers and architects, backed by the world's leading repository of code publications.

FREE CODE—LATEST EDITION

Most new individual members receive a free code from the latest edition of the International Codes®. New corporate and governmental members receive one set of major International Codes (Building, Residential, Fire, Fuel Gas, Mechanical, Plumbing, Private Sewage Disposal).

FREE CODE MONOGRAPHS

Code monographs and other materials on proposed International Code revisions are provided free to ICC members upon request.

PROFESSIONAL DEVELOPMENT

Receive Member Discounts for on-site training, institutes, symposiums, audio virtual seminars, and on-line training! ICC delivers educational programs that enable members to transition to the I-Codes®, interpret and enforce codes, perform plan reviews, design and build safe structures, and perform administrative functions more effectively and with greater efficiency. Members also enjoy special educational offerings that provide a forum to learn about and discuss current and emerging issues that affect the building industry.

ENHANCE YOUR CAREER

ICC keeps you current on the latest building codes, methods, and materials. Our conferences, job postings, and educational programs can also help you advance your career.

CODE NEWS

ICC members have the inside track for code news and industry updates via e-mails, newsletters, conferences, chapter meetings, networking, and the ICC website (www.iccsafe.org). Obtain code opinions, reports, adoption updates, and more. Without exception, ICC is your number one source for the very latest code and safety standards information.

MEMBER RECOGNITION

Improve your standing and prestige among your peers. ICC member cards, wall certificates, and logo decals identify your commitment to the community and to the safety of people worldwide.

ICC NETWORKING

Take advantage of exciting new opportunities to network with colleagues, future employers, potential business partners, industry experts, and more than 50,000 ICC members. ICC also has over 300 chapters across North America and around the globe to help you stay informed on local events, to consult with other professionals, and to enhance your reputation in the local community.

JOIN NOW! 1-888-422-7233, x33804 | www.iccsafe.org/membership

ICC INTERNATIONAL CODE COUNCIL

People Helping People Build a Safer World™

09-01530